Wetland Management and Sustainable Livelihoods in

In this book the authors argue for a paradigm shift in the way African wetlands are considered. Current policies and wetland management are too frequently underpinned by a perspective that views agriculture simply as a threat and disregards its important contribution to livelihoods. In rural areas where people are entrenched in poverty, wetlands (in particular wetland agriculture) have a critical role to play in supporting and developing peoples' livelihoods. Furthermore, as populations rise and climate change takes grip they will be increasingly important.

The authors argue that an approach to wetland management that is much more people focused is required. That is an approach that instead of being concerned primarily with environmental outcomes is centred on livelihood outcomes supported by the sustainable use of natural wetland resources.

The authors stress the need for Integrated Water Resource Management and landscape approaches to ensure sustainable use of wetlands throughout a river catchment and the need for wetland management interventions to engage with a wide range of stakeholders. They also assess the feasibility of creating incentives and value in wetlands to support sustainable use. Drawing on nine empirical case studies, this book highlights the different ways in which sustainable use of wetlands has been sought, each case focusing on specific issues about wetlands, agriculture and livelihoods.

Adrian Wood is Professor of Sustainability at the University of Huddersfield School of Business, and Director of Wetland Action.

Alan Dixon is a Senior Lecturer in the Institute of Science and the Environment, University of Worcester, UK.

Matthew McCartney is a Principal Researcher with the International Water Management Institute (IWMI), currently based in Lao PDR but until recently based in Ethiopia.

Wetland Management and Sustainable Livelihoods in Africa

Edited by Adrian Wood, Alan Dixon and Matthew McCartney

Taylor & Francis Group

LONDON AND NEW YORK

from Routledge

First edition published 2013
by Routledge
2 Park Square, Milton Park, Abingdon, Oxon, OX14 4RN

Simultaneously published in the USA and Canada
by Routledge
711 Third Avenue, New York, NY 10017

Routledge is an imprint of the Taylor & Francis Group, an informa business

British Library Cataloguing in Publication Data
A catalogue record for this book is available from the British Library

Library of Congress Cataloging-in-Publication Data
Wetland management and sustainable livelihoods in Africa / edited
by Adrian Wood, Alan Dixon and Matthew McCartney. – 1st ed.
p. cm.
Includes bibliographical references and index.
ISBN 978-1-84971-411-2 (hbk) – ISBN 978-1-84971-412-9 (pbk)
– ISBN 978-0-203-12869-5 (ebk) 1. Wetland management–Africa. 2.
Wetlands–Economic aspects–Africa. 3. Wetland agriculture–Africa.
4. Sustainable development–Africa. 5. Ecosystem services–Africa. I.
Wood, Adrian P. II. Dixon, Alan B. III. McCartney, Matthew P.
TC515.W48 2013
333.918153096–dc23
2012048134

ISBN13: 978-1-84971-411-2 (hbk)
ISBN13: 978-1-84971-412-9 (pbk)
ISBN13: 978-0-203-12869-5 (ebk)

Typeset in Baskerville by
GreenGate Publishing Services, Tonbridge, Kent

Printed and bound by CPI Group (UK) Ltd, Croydon, CR0 4YY

Contents

Illustrations

Figures

Tables

Boxes

Notes on contributors

Robert Bagyenda, a wetland resource use planner, is a Programme Officer for Water and Wetlands at the IUCN Uganda office in Kampala. Prior to this, he established and headed the Environment Office in the greater Soroti District in eastern Uganda, and worked for ten years as a Regional Wetlands Coordinator in the Wetlands Management Department of Uganda's Ministry of Water and Environment. He was also a visiting lecturer in environment, water and wetland resource management at Makerere University and Mbarara University of Science and Technology in Uganda. His efforts on wetlands have concentrated on wetland management planning and enhancing wetland livelihood benefits as a catalyst to wetland biodiversity conservation. He was also the National Project Coordinator for the UNDP-GEF-funded COBWEB project from which some lessons were drawn for this publication.

Alan Dixon is a Senior Lecturer in the Institute of Science and the Environment at the University of Worcester, and a Research Associate of Wetland Action. As a human ecologist, his research has focused on human–environment interactions in wetland systems in Africa, particularly the role of local knowledge and institutional arrangements in supporting sustainable livelihoods. He has worked extensively in Ethiopia since 1996 on a range of research projects in collaboration with local NGOs, and contributed to the FAO's recent Guidelines on Agriculture and Wetlands Interactions (GAWI) project.

Afework Hailu is Executive Director of the Ethio-Wetlands and Natural Resources Association (EWNRA), an Ethiopian Resident Charity Organization. Ato Afework has been involved in wetlands research and advocacy in Ethiopia since the late 1990s, and previously held positions in the Ethiopian Ministry of Agriculture and the Ministry of Coffee and Tea Development. Currently he is actively engaged in the execution of projects and programmes that promote sustainable environmental management, particularly in wetland environments, with full participation of local communities.

Daniel Jamu, an Ecosystem Ecologist and Aquaculturist by profession, is Senior Scientist at the WorldFish Center Office in Malawi. He has managed and implemented aquaculture development and fisheries ecology research projects in Malawi, Mozambique and Zambia. His work on fisheries ecosystems has focused on the interactions between lake catchment management, fisheries production and livelihoods of fisheries dependent communities.

Paul Kiepe is a land and water management specialist by profession. He is currently the AfricaRice Regional Representative for eastern and southern Africa and is based in Dar es Salaam, Tanzania. He has worked in Africa for the last 25 years, of which the last ten for the Africa Rice Center (AfricaRice), which is a pan-African research-for-development organisation and a member of the CGIAR Consortium. From 2002 until 2006 he was Coordinator of the Inland Valley Consortium (IVC), which was hosted by AfricaRice. IVC was a so-called systemwide ecoregional CGIAR program focusing on integrated inland valley development in 12 West African countries. Typical activities were either biophysical, such as small-scale water management, integrated soil fertility management and integrated weed management, as well as social-economical, such as value chain analysis, multi-stakeholder platforms, adoption and impact.

Donovan Kotze, an Honorary Research Fellow at the University of KwaZulu-Natal, South Africa, has a particular interest in wetland ecology and the sustainable use of wetland systems. Over the last 20 years he has worked in a variety of southern African wetlands under many different land uses, from communal, traditional use to intensively used private land. He is especially interested in wetlands within an agricultural context, and collaborates with many different government and non-government organisations, particularly the Mondi Wetlands Programme. He has also worked extensively on developing and applying methods for the assessment of wetland ecological condition and the provision of ecosystem services, used for a variety of purposes, including rehabilitation and management planning and environmental impact assessments.

Matthew McCartney is a principal researcher specializing in water resources and wetland and hydro-ecological studies at the International Water Management Institute (IWMI). Until the end of 2011 he was based in IWMI's office in Addis Ababa, Ethiopia and is now in Vientiane, Lao PDR. His experience stems from participation in a wide range of research and applied projects, many of which relate to wetlands and wetland livelihoods and agriculture. He has been IWMI's representative on the Ramsar Science and Technical Review Panel (STRP) since 2007.

Barbara Nakangu is the Head of the IUCN Uganda office in Kampala and is charged with coordinating an array of programs that cut across wetlands, forests and dry lands. The overall aim of the IUCN programme is to develop evidence-based solutions to the development challenges today that catalyse the dual function of conserving nature and securing livelihoods. She has background in Agricultural Economics and Natural Resources Management and has worked on issues related to these sectors for over 12 years. Her work has influenced policy and practice at different levels in the country. Her career includes working at the Uganda National Council for Science and Technology; a Uganda National NGO known as FIT Uganda and the IUCN where she has spent eight years.

Sharon Pollard is the director of AWARD (the Association for Water & Rural Development), a research and development non-profit organisation that focuses on the integration of research and implementation for water resource management in southern Africa. Trained as a freshwater ecologist her interests have expanded to include social and policy issues framed by approaches that examine sustainability through the lens of systems thinking and social learning. Her work has focused on a broad range of topics including dam-impacted communities in the Amazon and south-east Asia, and freshwater sustainability in southern Africa through Integrated Water Resources Management, transboundary water-sharing arrangements and research into wetlands, livelihoods and governance.

Lisa-Maria Rebelo, a remote sensing specialist by profession, is a researcher at the International Water Management Institute, part of the Consultative Group for International Agricultural Research. Her research is focused on the provision of spatial information for wetland inventory, assessment and monitoring, with a focus on wetland ecosystems that are used for agriculture in Africa.

Jonne Rodenburg has a PhD in Production Ecology and Resource Management and an MSc in Tropical Agronomy, both obtained at Wageningen University in the Netherlands. He is currently employed by the Africa Rice Center (AfricaRice, a CGIAR consortium member) as Rice Agronomist for East and southern Africa. He has worked in Africa for ten years, of which the last eight for AfricaRice, based in Mali (2004), Benin (2005–2007), Senegal (2007–2009) and Tanzania (2009–present). He worked on a diversity of rice research topics ranging from integrated weed management and Good Agricultural Practices, to varietal weed competitiveness and parasitic weed resistance, and from elucidating biology and ecology of parasitic weeds to developing socially acceptable weed management strategies and water-saving production methods. In 2010 he won the CGIAR Science Award for Promising Young Scientist.

Tilahun Semu is currently the Field Office Co-ordinator for the Ethio-Wetlands and Natural Resources Association (EWNRA) in Ilu Aba Bora. For the last 25 years he has been working in agricultural extension and natural resource management, having held positions previously in the Ethiopian Ministry of Agriculture. His current work for the EWNRA focuses on sustainable natural resource management, specifically on watershed-based natural resource wetland management.

Katherine A. Snyder is a social anthropologist by training and is currently a senior social scientist at the International Water Management Institute in Addis Ababa, Ethiopia. She has worked in Africa for more than 20 years, focusing mainly on Tanzania, Kenya, Malawi and Ethiopia. Her work has analysed the interface between livelihoods, natural resources, development and governance in rural communities. She has written on gender, politics, religion and community-based natural resource management.

Adamu I. Tanko, a development geographer, is the Dean, Faculty of Social and Management Sciences at Bayero University Kano, Nigeria. He has managed field research in northern Nigeria and beyond for about 20 years and led a number of funded research projects. His main areas of research are environment and development, and irrigation management. He has published over 40 articles on various topics in the social sciences and written four books. He has won a number of awards including the Ron Lister Fellowship at the University of Otago, New Zealand.

Patrick Thawe, a rural development practitioner, is a project coordinator for the Malawi Enterprise Zone Association (MALEZA). He has worked with MALEZA for more than seven years coordinating different livelihoods and climate change resilience projects and sustainable wetland management. He managed the Striking a Balance (SAB) wetland demonstration project supported by Wetland Action (WA) and Wetlands International (WI) in Malawi.

Derick du Toit is an ecologist specialising in environmental learning in multi-stakeholder environments. He is currently project manager for the Association for Water & Rural Development (AWARD), a non-profit, research-based organisation. He has worked in South Africa, Namibia, Mozambique and Botswana as both an ecologist and education specialist, including work on national education reform in three southern African countries, and has authored textbooks, teacher manuals and research papers. His current interests include working in rural contexts with natural resource management practices.

Adrian Wood, a human geographer by profession, is Professor of Sustainability at the University of Huddersfield School of Business. He was formerly Director of the Centre for Wetlands, Environment and Livelihoods at the same university. He has worked in Africa for more than 30 years and has managed wetland research projects in Ethiopia, Malawi and Zambia. His work on wetlands has focused on the sustainable management of these areas by communities for diverse livelihood benefits, especially food and water. Along with a Dutch colleague, Rob de Rooij, he established Wetland Action as a not-for-profit organisation to support this sustainable use approach to wetland management, through direct project implementation, by disseminating research findings and through support to international, government and non-governmental organisations.

Acknowledgements

The editors and authors wish to record their sincere appreciation to all the wetland-using communities who gave their time to inform the studies on which this book is based. It is hoped that we have captured their experience and views and that this book will make a contribution to improving their livelihoods and the state of wetlands in Africa.

The roles are recognised of several NGOs, international agencies and government organisations with whom the authors have worked and who have funded many of the activities reported here.

The authors wish to record their appreciation of the support from their employers who have facilitated and encouraged the production of their chapters and the book. Support for the involvement of the editors is specifically acknowledged:

- Alan Dixon – the University of Worcester's Project Leave Scheme;
- Matthew McCartney – the CGIAR Research Program on Water, Land and Ecosystems, and the International Water Management Institute (IWMI) which is a member of the CGIAR consortium and leads this program; and
- Adrian Wood – the University of Huddersfield and Wetland Action.

The editors would also like to express their appreciation of financial support from several different organisations which over the last 15 years have supported their research in different African countries and which have contributed to the development of ideas for this book. These include the European Union's Tropical Environment Budget Line for a collaborative study in Ethiopia, the Leverhulme Trust for work in Zambia, the UK Economic and Social Research Council and the British Academy for work in Ethiopia, the FAO Netherlands Partnership Program for work in Zambia and Tanzania, the Global Environment Facility (GEF) for work throughout Southern Africa and the CGIAR Challenge Program for Water and Food (CPWF) for work in the Limpopo basin.

The financial support from the IWMI and Wetland Action towards the publication of this book is gratefully acknowledged.

The editors would like to thank the wetland and livelihood specialists who commented on the first chapter of this book for giving their time and constructive comments. These were Gerardo van Halsema of Wageningen University, the Netherlands, Declan Conway of the University of East Anglia, UK, and Max Finlayson of Charles Sturt University, Australia.

In the production of this work Sian Evans' formatting of the text is greatly appreciated, while Tim Hardwick and Ashley Wright of Earthscan and Karen Wallace of GreenGate Publishing Services are thanked for their assistance throughout.

The editors would also like to thank the authors for their positive responses to their comments and various requests in the process of finalising this book.

Finally, we would reiterate that it is the time, efforts and experience of wetland users, mostly farmers, which have made this book possible and for which we are all grateful.

Acronyms and abbreviations

ACRU	Agricultural Catchments Research Unit
ADB	African Development Bank
ADC	area development committee
ADF	African Development Fund
ARICA	Advanced Rice for Africa
BIG	Bumbwisudi Irrigation Group
BMU	Beach Management Unit
BVC	Beach Village Committee
CA	Comprehensive Assessment of Water Management in Agriculture
CADP	Commercial Agricultural Development Project
CBD	Convention on Biological Diversity
CBNRM	community-based natural resource management
CBO	community-based organisation
CCA	Community Conserved Area
CDD	Community-Driven Development
CEC	cation exchange capacity
CFC	Common Fund for Commodities
CGIAR	Consultative Group on International Agricultural Research
CIRAD	Centre de Coopération Internationale en Recherche Agronomique
CLUSA	Cooperative League of the USA
COBWEB	Community-Based Wetlands Biodiversity conservation project
COP	Conference of the Contracting Parties
CORAF	West and Central African Council for Agricultural Research and Development
CPR	common property resource
Danida	Danish International Development Agency
DFID	Department for International Development
DIARPA	Diagnostic Rapide de Pré-aménagement
DPSIR	Drivers-Pressures-State Changes-Impacts-Responses
EIA	Environmental Impact Assessment
ESS	ecosystem services

EWNRA	Ethio-Wetlands and Natural Resources Association
EWRP	Ethiopian Wetlands Research Programme
FAO	Food and Agriculture Organization of the United Nations
FLA	functional landscape approach
FSP	Farmer Support Programme
GAWI	Guidelines on Agriculture and Wetlands Interaction
GDP	gross domestic product
GEF	Global Environment Facility
GIS	Geographical Information System
GMA	Game Management Area
GOM	Government of Malawi
HJKYB	Hadejia-Jama'are-Komadugu-Yobe basin
HJKYBTF	Hadejia-Jama'are-Komadugu-Yobe basin Trust Fund
HJRBDA	Hadejia-Jama'are River Basin Development Authority
HNWCP	Hadejia-Nguru Wetlands Conservation Project
HVIP	Hadejia Valley Irrigation Project
IAD	Institutional Analysis and Development
ICBP	International Council for Bird Preservation
ICDP	Integrated Conservation and Development Project
IDS	Institute of Development Studies
IFDC	International Fertilizer Development Center
IIED	International Institute for Environment and Development
IITA	International Institute of Tropical Agriculture
IK	indigenous knowledge
ILRI	International Livestock Research Institute
INERA	Institut de l'Environnement et de Recherches Agricoles
IPCC	Intergovernmental Panel on Climate Change
IRM	Integrated Rice Management
ITM	Integrated Transect Method
IUCN	International Union for the Conservation of Nature
IVC	Inland Valley Consortium
IWMI	International Water Management Institute
IWRB	International Waterfowl and Wetlands Research Bureau
JEWEL	Jigawa Enhancement of Wetlands and Livelihood
JICA	Japan International Cooperation Agency
KAP	Knowledge, Attitudes and Practices
KDC	Kasungu District Council
KDPD	Kasungu District Planning Department
LCBCCAP	Lake Chilwa Basin Climate Change Adaptation Programme
LVDP	Luano Valley Development Program
LWMI	local wetland management institution
MA	Millennium Ecosystem Assessment
MALEZA	Malawi Enterprise Zone Association
MARD	Ministry of Agriculture and Rural Development
MCDM	multi-criteria decision making

MDG	Millennium Development Goal
MFPED	Ministry of Finance, Planning and Economic Development
MNR	Ministry of Natural Resources
MoA	Ministry of Agriculture
MSP	multi-stakeholder platform
MWLE	Ministry of Water, Lands and Environment
NARES	national agricultural research and extension systems
NEMA	National Environment Management Authority
NERICA	New Rice for Africa
NFDP	National Fadama Development Programme
NGO	non-governmental organisation
NHI	National Heritage Institute
NIWRMC	Nigeria Integrated Water Resources Management Commission
NRM	natural resource management
NU	Nature Uganda
PA	Protected Area
PAF	Poverty Alleviation Fund
PES	payment for environmental services
PLAR	Participatory Learning and Action Research
PRA	Participatory Rural Appraisal
RAP	Realizing the Agricultural Potential of Inland Valley Lowlands in Sub-Saharan Africa
ROSA	Regional Office for Southern Africa (IUCN)
ROU	Republic of Uganda
SAB	Striking a Balance
SES	socio-ecological systems
SLA	Sustainable Livelihoods Approach
SLF	Sustainable Livelihoods Framework
SOM	soil organic matter
SRTM	Shuttle Radar Topography Mission
SSA	sub-Saharan Africa
TA	traditional authority
TF	Trust Fund
UBOS	Uganda Bureau of Statistics
UNDP	United Nations Development Program
UNEP	United Nations Environment Programme
UNICEF	United Nations Children's Fund
UWS	Uganda Wildlife Society
VDC	village development committee
VNRMC	Village Natural Resource Management Committee
WA	Wetland Action
WAIVIS	The West Africa Inland Valley Information System
WARDA	West African Rice Development Association
WARFSA	Water Research Fund for Southern Africa
WCD	World Commission on Dams

WI	Wetlands International
WID	Wetlands Inspection Division
WMD	Wetlands Management Department
WPRP	Wetlands and Poverty Reduction Project
WRI	World Resources Institute
WSSP	Wetland Sector Strategic Plan
WWF	World Wide Fund for Nature
WWP	Working Wetland Potential
ZAWA	Zambia Wildlife Authority

1 People-centred wetland management

Adrian Wood, Alan Dixon and Matthew McCartney

Introduction

Wetlands have played a critical role in the livelihoods of people in Africa for millennia, not least because they have been sources of food and water for people living in often dry and semi-arid environments (Scoones, 1991). Indeed, much has been made in the literature of the last 30 years of the capacity of wetlands to provide a diverse range of functions and services that have supported people, ecological systems and the physical environment (Maltby, 1986; Dugan, 1990; Davis, 1994; Barbier *et al.*, 1997; MA, 2005). Yet as explored throughout this book, the functional relationship between wetlands, people and livelihoods, and indeed the very survival of people, has tended to be neglected in the global hegemonic discourse of wetland management that has largely been driven by environmental concerns (Matthews, 1993; Melamed *et al.*, 2012). While the effect of this discourse in Africa has been a growing awareness of the environmental importance of wetlands among policy-makers, a persistent view has developed that local people constitute the principal agents of wetland destruction; for example, the transformation of wetlands by drainage for subsistence agriculture has, and continues to be, regarded in many places as unacceptable (see Chapter 7). Consequently, as will be discussed later in this chapter, many wetland policy and management initiatives have sought to largely exclude potentially 'destructive' farmers and pastoralists.

This book, however, argues that such policies are no longer acceptable or relevant in light of the challenges for human development and livelihood security facing Africa in the twenty-first century (FAO, 2011; UNDP, 2012a). Rather, it is argued here that peoples' use of wetlands for multiple sustainable benefits, of an economic, social and environmental nature, must be the main focus (Howard *et al.*, 2009). The majority of Africa's population (estimated to be around 1 billion) continue to live in rural areas where a life of smallholder subsistence agriculture, and a lack of access to basic needs, such as food and water, have entrenched many people in poverty (Binns *et al.*, 2011). Despite some progress towards achieving the Millennium Development Goals (MDGs), recent statistics suggest that food insecurity and under-nourishment continue to rise and this will remain a major challenge not least because of population growth (FAO, 2006, 2011; UNDP, 2012b). It is estimated that 239 million people are undernourished in Sub-Saharan Africa (FAO, 2011) and around 340 million people across the continent (the majority of whom live in rural areas) continue to lack access to safe drinking water (UNICEF, 2006). Reports from the Intergovernmental Panel on Climate Change (IPCC) and more recent scientific studies suggest that climate change across the continent is also likely to compromise food security and agricultural livelihoods due to changes in growing seasons, increased rainfall variability and water shortages (Boko *et al.*, 2007; Lobell *et al.*, 2008; Muller, 2009).

These are the issues that must be addressed, and as this book argues emphatically, wetlands have a critical role to play in supporting *and* developing peoples' livelihoods, reducing poverty, improving food security and, in the wider context, contributing towards sustainable development. We do not advocate the indiscriminate exploitation of wetlands but rather a balanced approach that seeks to optimise benefits for poor rural populations and simultaneously safeguards vital ecosystem services (ESS). In this book, we present a number of case studies from around the continent which: a) illustrate the contribution that wetland benefits, especially agriculture, can make to peoples' livelihoods and well-being, and b) incorporate ideas, concepts and initiatives that have sought to facilitate win–win outcomes in wetlands for both the environment and development. In this way, the book seeks to reconceptualise wetlands as natural resources within the discourse of sustainable development, and hence reposition the wider wetland management debate away from environmental management with development outcomes, to one of livelihood development based on the sustainable use of these resources which inherently ensures positive environmental outcomes. Consequently, the case studies draw, in particular, upon concepts and ideas that have emerged from the development and natural resource management (NRM) literature, such as community-based participatory approaches, common pool resources, social and ecological resilience, integrated water resource management and catchment planning, and critically the concept of a sustainable livelihood itself. These are applied primarily to small inland wetlands in different parts of the African continent where small-scale agriculture is a major contributor to rural livelihoods.

In this chapter we provide an overview of the wetlands discourse of the last 60 years, as well as identifying other areas of thinking that can contribute to taking this discussion forward. The nature of wetlands and the ESS they provide, especially as contributions to African livelihoods, are reviewed in the first part of the chapter. The wetland discourse and the relevant thinking from the development agenda are reviewed in the next two sections, exploring how different schools of thought and concepts have contributed to the thinking on wetlands, and can inform current wetland debates. The chapter concludes with a focus on recent thinking about wetlands and proposes a framework for a 'people-centred approach for analysing the management of small inland wetlands in Africa'. The aim of this wide-ranging introduction is to ensure readers of different experience and interests are introduced to the diverse thinking and ideas that have been, and could be, relevant to ensuring that wetland management contributes to sustainable livelihood development in Africa.

Wetlands and ESS

Wetlands are diverse environments, both spatially and temporally, and also in terms of their physical size, ecology, hydrology and geomorphology. Whilst a myriad of literature presents and debates the definition and characterization of wetlands in different environments, along with their associated functions (Dugan, 1990; Roggeri, 1995; MA, 2005; Maltby and Barker, 2009), the most widely accepted scientific definition for conservation and planning purposes continues to be that established by the Ramsar Convention in 1971. This definition embraces the diversity of wetlands by grouping together a wide variety of landscape units whose ecosystems share the fundamental characteristic of being strongly influenced by water:

> areas of marsh, fen, peatland or water, whether natural or artificial, permanent or temporary, with water that is static or flowing, fresh, brackish or salt, including areas of marine water the depth of which at low tide does not exceed six metres.
>
> (Davis, 1994: 3)

Subsequently the Ramsar Bureau developed a classification of 42 distinct wetland types (Davis, 1994) although it later simplified these into five broad categories:

- marine (coastal wetlands including coastal lagoons, rocky shores and coral reefs);
- estuarine (including deltas, tidal marshes and mangrove swamps);
- lacustrine (wetlands associated with lakes);
- riverine (wetlands along rivers and streams);
- palustrine (meaning 'marshy' – marshes, swamps and bogs).

(Ramsar, 2009)

All of these types of wetland are found throughout Africa, and estimates suggest that they constitute somewhere between 1 per cent and 16 per cent of the total land area of African countries depending upon whether the larger lakes are included within this classification (Hughes, 1996; Schuyt, 2005; Rebelo *et al.*, 2010). This is likely to be an underestimate, however, since the data on wetlands is far from comprehensive due to inconsistencies in the wetland terminology used, the neglect of small seasonal wetlands and the logistical constraints of carrying out surveys (Denny, 2001).

The occurrence of wetlands (Figure 1.1) reflects the variation in climate and geomorphology, and Denny (1993) identifies two broad physiographic units: 'low Africa', situated to the north and west, which is composed of sedimentary basins and upland plains below 600 m and favours the formation of floodplain wetlands; in contrast, 'high Africa' to the south and east, which can be characterised by the results of tectonic activity, includes extensive mountainous areas, deep valleys and highland plateaux. Within the latter, lakes and swamps tend to be the more abundant wetland features.

Much has been written about the value and functions of wetlands (Dugan, 1990; Barbier, 1993; Roggeri, 1995; MA, 2005; Maltby, 2009; Maltby and Barker, 2009), and as greater understanding has emerged of the role wetlands play in ecological and hydrological cycles, there has been a concurrent

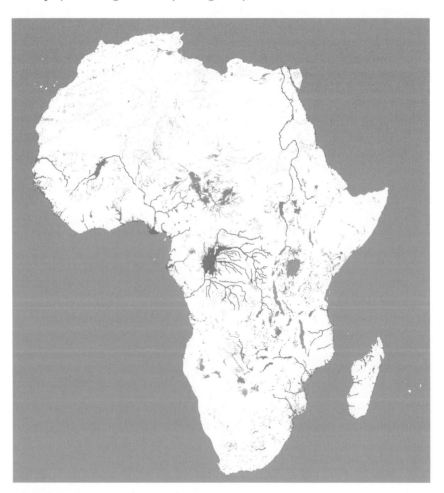

Figure 1.1 Distribution of wetlands in Africa

Source: adapted from MA, 2005

shift in attitudes away from wetlands being viewed as unproductive wastelands, to one of wetlands as multifunctional resources (Figure 1.2).

The multifunctional nature of wetlands has most recently been concep-
tualised in the Millennium Ecosystem Assessment's (MA) classification of
'wetland ecosystem services' which stresses how they contribute to human
well-being and poverty alleviation (MA, 2005: 1). The range of ESS that wetlands can provide, either in their natural state or when partially transformed,
are highlighted in Table 1.1. Many ESS, especially provisioning ones, have
direct economic value to people, while regulating and support services in
general help maintain environmental functions that are of benefit to communities. Wetlands also provide ESS which have other, cultural, values for
communities, especially with respect to spiritual, recreational and aesthetic
interests. In addition, wetlands have biodiversity values that have been a key
element in the discourse about the management of these areas. Different
societies under different socio-economic, cultural and development conditions value these ESS differently and this adds to the diversity that needs to
be recognized when discussing wetlands.

The MA also makes it clear that not all wetlands support the full range
of ESS. Specific services are associated with particular types of wetland in
specific ecological and geographical settings as well as the ecological condition of the wetland. Because of the complexity of natural systems it is often
difficult to predict the exact nature and the magnitude of services that any
given wetland provides. With respect to hydrological regulating services,
research has shown that these wetland services are not as widespread as
once thought (Bullock and Acreman, 2003).

Figure 1.2 Multifunctional seasonal wetland use in central Malawi

Source: Adrian Wood

Table 1.1 Ecosystem services provided by, or derived from, wetlands

Services	Comments and examples
Provisioning	
Food	Production of fish, wild game, fruits and grains
Freshwater	Storage and retention of water for domestic, industrial and agricultural use
Fibre and fuel	Production of logs, fuelwood, peat and fodder
Biochemical	Extraction of medicines and other materials from biota
Genetic material	Genes for resistance to plant pathogens, ornamental species, etc.
Regulating	
Climate regulation	Source and sink for greenhouse gases; influence local and regional temperature, precipitation and other climate processes
Water regulation (hydrological flows)	Groundwater recharge/discharge
Water purification and waste treatment	Retention, recovery and removal of excess nutrients and other pollutants
Erosion regulation	Retention of soils and sediments
Natural hazard regulation	Flood control and storm protection
Pollination	Habitat for pollinators
Cultural	
Spiritual and inspirational	Source of inspiration; many religions attach spiritual and religious values to aspects of wetland ecosystems
Recreational	Opportunities for recreational activities
Aesthetic	Many people find beauty or aesthetic value in aspects of wetland ecosystems
Educational	Opportunities for formal and informal education and training
Supporting	
Soil formation	Sediment retention and accumulation of organic matter
Nutrient cycling	Storage, recycling, processing and acquisition of nutrients

Source: adapted from MA, 2005

Wetlands, livelihoods and sustainability in Africa

Wetlands and livelihoods

In general, it is in their capacity to provide what the MA (2005) calls 'provisioning services', that African wetlands arguably make the most important direct contribution to peoples' livelihoods. In particular, wetlands have been a critical source of food, fresh water, fibre and fuel, in many of the most marginal areas of Africa for centuries (Trapnell and Clothier, 1937; Gluckman, 1941; Hollis, 1990; Scoones, 1991). This relationship between communities and wetlands continues today as provisioning services are increasingly developed, and as wetlands play ever more important roles in livelihood diversification in the face of challenges to traditional livelihoods emerging from population growth and climate change. In northern Nigeria, for example, seasonally flooded, low-lying areas known as fadamas support the livelihoods of hundreds of thousands of smallholder farmers (Dan-Azumi, 2010: Chapter 9). Similarly throughout central and southern Africa, seasonally inundated dambos are common landforms that provide a reservoir of soil moisture during the dry season and hence represent valuable agricultural resources (Turner, 1986: Chapters 3 and 6). Across eastern and southern Africa, large riverine floodplains, lakes, and permanent and seasonal swamps support livelihoods such as fishing, agriculture, pastoralism and craft production (Crafter *et al.*, 1992; Kamukala and Crafter, 1993; Abebe and Geheb, 2003; FAO, 2004). The role of wetlands in supporting major concentrations of people is clear, both historically – as in the lower Nile (Howell and Allan, 1994) and on the Upper Zambezi floodplain (Gluckman, 1941) – and more recently around many irrigation schemes (Chambers, 1969).

Some assessments of the present economic contribution of wetlands confirm the continued importance of these areas. In Zambia the total use value of wetlands (with fish production and floodplain recession agriculture accounting for the main share) was estimated to be the equivalent of approximately 5 per cent of Zambia's gross domestic product (GDP) in 1990 (Seyam *et al.*, 2001). Looking at the Zambezi river basin as a whole, across eight countries, the economic value of wetlands in terms of crops alone was estimated to be close to US$50 million a year (UNEP, 2006). In addition, the value of wetland fisheries in the basin is estimated to be US$80 million a year, while its floodplain grasslands support livestock production is valued at over US$70 million annually.

More detailed studies have found that wetlands contribute significantly to the livelihoods of households living near them. In the Kilombero valley in Tanzania, the contribution of wetland cultivation to cash income was 66 per cent of the approximately US$518 cash income per household per year, with poor households getting 80 per cent of this income from wetland cultivation, while the intermediate and better-off households obtained 70 per cent and 48 per cent of their total cash income respectively from wetland cultivation (McCartney and van Koppen, 2004: Chapter 2). Similar findings come from other studies that have shown a wide range in household income generated from wetlands, especially through cropping. In the Ga-Mampa wetland in South Africa the average annual value of wetland cultivation per household was estimated at US$93 (Adekola *et al.*, 2008); in the Nakivubo urban wetland in Kamapala, Uganda it was US$300 per household (Emerton, 2005); in the Barotse floodplain in Zambia US$109 per household (Turpie *et al.*, 1999); in the Lower Shire, Malawi US$363 per household (Turpie *et al.*, 1999); and in the Chipala Ibenga wetland, Zambia, from US$19 to US$107 per household (Masiyandima *et al.*, 2004). Household dependence on wetlands and hence demands on wetland resources are, however, highly site specific. While they are partly influenced by the nature of the resources in the wetland, they are also subject to access to markets and changing demands in these markets. Furthermore, socio-economic differentiation leads to significant disparity in the agricultural utilization of the wetlands and the benefits derived from them (see Chapters 2 and 3).

In recent years the use of wetlands in Africa, especially for agriculture, has increased to the extent that wetlands have now become a 'new agricultural frontier' in the continent (Figure 1.3) (Dixon and Wood, 2003: 119). The reasons for this are diverse but include: rural population growth and the consequent shortage of farmland; the degradation of upland or rain-fed fields; a reduction in the quantity and reliability of rain-fed harvests; increased needs for earning cash; and the development of new technologies to use these areas, such as treadle pumps (Inocencio *et al.*, 2003). The demand for the livelihood benefits from wetlands is likely to increase further as the impacts

Figure 1.3 Rice harvesting in an inland valley wetland in Burkina Faso

Source: Jonne Rodenburg (AfricaRice)

of climate change increase, the search for water intensifies in some areas and plots where secure harvests can be obtained become increasingly sought-after. Indeed wetlands will become a critical area in the process of adapting to climate change and mitigating its impact – a process that has arguably already started in many areas (Boko *et al.*, 2007; Osbahr *et al.*, 2008).

With these challenges, wetlands are becoming an increasingly valuable multiple use resource in Africa, and are increasingly seen as a safety net for many, especially the poor, to help them cope with the struggle for survival. However, wetlands are also being developed in some places by innovative farmers, often with pumps or other investments, to meet market demands for vegetables (Figure 1.4) and so are becoming a source of capitalist enterprise development (Woodhouse *et al.*, 2000). A critical challenge in both these situations is to ensure that wetlands are used and managed in a manner that does not degrade the natural resource base from which the various livelihood benefits and ESS are derived and sustained. This is particularly pertinent given that numerous studies have drawn attention to the ways in which wetlands have been degraded as a result of inappropriate management and utilization practices (Roggeri, 1995).

Figure 1.4 Vegetable cultivation in a partially drained, permanent wetland in Malawi

Source: Adrian Wood

In view of this wide-ranging use of wetlands by rural communities in the continent, and their growing importance, it is clear that people and their management of wetlands are an essential part of the wetlands discourse. The relationship between people and wetlands must be central to the development of wetland policies and wetland management approaches so that they can ensure sustainable livelihoods from these areas.

Sustainability and wetlands

There is widespread evidence in Africa of abandoned and degraded wetlands, the result in most cases of overdrainage and/or erosion that has lowered or disrupted water tables (Roggeri, 1995; Schuyt, 2005). The loss of wetlands through inappropriate management in these areas, and also in adjoining catchments (McFarlane and Whitlow, 1990), draws attention to the fact that the activities of communities in their wetlands need to be sensitive to the environmental processes that are on-going, not least because, in most cases, wetlands are temporary features in the long-term processes of erosion and succession.

It is imperative, therefore, that wetland use strategies integrate the three key principles of sustainability, i.e. they should be:

- *Environmentally sustainable*, which means facilitating a wetland use strategy or management plan (which may include agriculture) whose development of provisioning or other sought-after ESS does not irreversibly degrade the wetland and rather maintains the other ESS – typically regulating and supporting – for the foreseeable future.
- *Economically sustainable*, which means facilitating wetland use that is economically attractive to users, for which there are clear economic development benefits that continue in the long term.
- *Socially sustainable*, which means that wetland management and utilization strategies are rooted within the local community, developed by local people and based on local knowledge. Wetland management should also be socially and economically equitable so as to avoid conflict which can undermine sustainable use.

While sustainability as a concept has been recognized in the wetland literature, and the wetland management policy agenda has sought to incorporate principles of sustainability into wetland management strategies since the early 1980s, the extent to which this process has been successful in terms of meeting both the development needs of local people and environmental concerns in an inclusive approach is debateable. Indeed there have been calls for a wider interpretation of sustainability in the discussion on wetlands and it is to the differences in that discourse that we now turn.

The wetlands discourse

The position taken throughout this book, that wetlands have a critical role to play in developing the livelihoods of people in Africa, especially the poor, is situated within the classic environment and development debates that have dominated the global discourse of NRM for many years, and especially with respect to wetlands. Two perspectives/contrasting positions on wetlands can be identified: wetland development and wetland conservation, with evidence of some convergence of thinking occurring over the past 20–30 years (Figure 1.5).

Wetland development

At one end of the wetland management spectrum has been a *utilitarian* and *development* view, which has encouraged the transformation and use of wetlands by people. Until interest in wetland conservation emerged during the 1930s the dominant view was one of wetlands as unhealthy, unproductive areas that could be more productive if their hydrological regime was manipulated, often by drainage. Certainly this was the case throughout much of Europe where wetland drainage over several centuries allowed large areas of land to be transformed and cultivated (Cook and Williamson, 1999). In Africa, however, agriculture in wetlands was widespread before the arrival of Europeans as evidenced in folk stories, the visible remains of agricultural practices (e.g. raised bed and furrows) and the literature from early European explorers (Cecchi, 1886; Trapnell and Clothier, 1937; Mharapara, 1994). In West Africa, despite constraints to cultivation (i.e. extensive weed growth, lack of water management technologies and prevalence of water related disease) wetlands were used for rice cultivation (Andriesse *et al.*, 1994). In southern Africa, crops such as *Clues esculentus* (tsenza), cucurbits, cocoyams and a variety of vegetables were traditionally grown by farmers in dambos (Mharapara, 1994), while in western Zambia intensive cultivation was practised along the edge of the Upper Zambezi floodplain (Gluckman, 1941).

During the colonial period, European farmers in southern Africa were also attracted to wetlands by the 'turf like' soils that were easily ploughed and the high moisture retention that allowed cropping in the dry winter months (Whitlow, 1990). The value of wetlands for agriculture was widely acknowledged by agricultural engineers with journal articles of the time discussing the merits but also the environmental risks associated with wetland agriculture:

> Vleis have considerable agricultural potential mainly by virtue of the fact that they remain wet far into the dry season or even throughout the dry season … The principal danger in vlei drainage is the development of a gully down the centre of the vlei. Such a gully will be uncontrolled drainage which may and often does completely dry out the vlei resulting

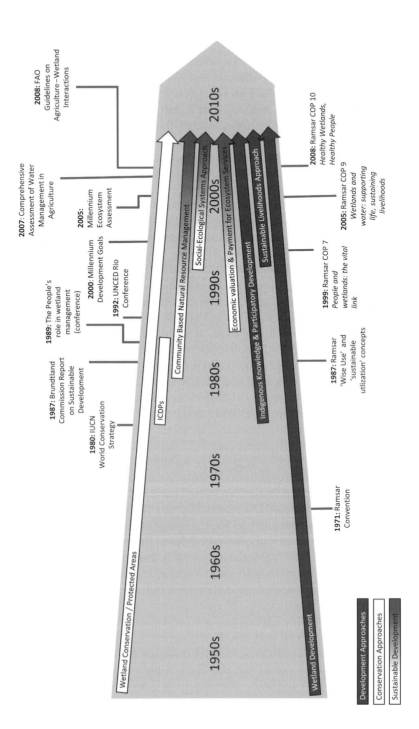

Figure 1.5 Convergence of thinking towards a people-centred sustainable wetland management approach

in destruction of its naturally good vegetative cover, the destruction by
oxidation of the high humus content of the soil and ending with further
erosion and ultimate destruction of what was once a valuable asset.

(Rattray *et al.*, 1953)

It is interesting that in Southern Rhodesia (now Zimbabwe) legislation was
introduced to prohibit the use of wetlands (dambos) for agriculture long
before conservation *per se* was a major concern. Although the concept of ESS
was unknown at the time, the Water Act of 1927 and the Natural Resources
Act of 1941 (revised in 1952) prohibited any cultivation along stream banks
and in dambos in order to protect another perceived service of these wetlands: the provision of water. Despite this legislation, cultivation of dambos
by settler farmers increased substantially through the 1930s and 1940s
(Whitlow, 1990), and it was only in the 1950s and 1960s that conservation
pressures resulted in a decrease in the cultivation of dambos. Despite there
being little evidence to support it, traditional small-scale farming of dambos was also condemned. For example, a Commission of Enquiry in 1939
denounced peasant cultivation in dambos, claiming that 'once pleasant valleys [were] transformed by this means to barren stretches of gaping dongas
[i.e. gulleys]' (McFarlane and Whitlow, 1990: 217). However, enforcement
of legislation was very variable, in some places farmers were moved off their
dambo gardens, whilst in others dambo cultivation continued (Bell and
Roberts, 1991).

During the last 50 years since independence in most African countries,
there has been a significant drive for the transformation of wetlands, especially for cultivation. This has come in part from government agencies that
see wetland cultivation as a critical way of achieving food security (Wood
et al., 2001) or producing export crops, but also from commercial com-
panies seeking to develop profitable enterprises. This has been despite
other conflicting government policies that have espoused the conservation
of wetlands. In addition, there is widespread evidence from across sub-
Saharan Africa (SSA) of individual farmers using wetlands for vegetable
growing in response to market opportunities (Woodhouse *et al.*, 2000) and
in other cases as a source of food during the hungry season (Figure 1.4)
(Ndiyoi *et al.*, 2009). Examples of this expansion of wetland cultivation
on a large scale can be found from around the continent including the
development of rice cultivation in the inland valleys of West Africa (see
Chapter 10), sugar estates in many countries (Temper, 2010), irrigated
cultivation of cotton in countries such as Ethiopia (Bondestam, 1974) and
the development of the dairy industry in Uganda (see Chapter 7).

Today, although many (mainly 'northern' academics) have argued
vociferously against the conversion of wetlands on the grounds that it is
not sustainable and the temporary benefits are outweighed by the loss of
wetland ESS (see Maltby and Turner, 1983; Dugan, 1990; Hollis, 1990), the
conversion of wetlands to agricultural land remains an attractive option

that is used by many farmers throughout Africa. Indeed, for most governments food security concerns are higher up the political agenda than biodiversity conservation and the potential contribution of wetland agriculture to poverty alleviation and development is now widely recognized (Ferreira, 1977; Perara, 1982; Millington *et al.*, 1985; Bell *et al.*, 1987; Daka, 1997; Makombe *et al.*, 2001; Kangalawe and Liwenga, 2005; McCartney *et al.*, 2005; McCartney *et al.*, 2010; Verhoven and Setter, 2010). In Zambia, the National Irrigation Policy of 2005 acknowledges the importance of dambos and wetlands as sources of water that smallholder and emergent farmers can utilize with low-cost irrigation technology. The potential of wetlands for agriculture is also appreciated by some local authorities. For instance, the district council in the vicinity of the Lukanga Swamp (a major wetland in the catchment of the Kafue River) has recently produced a development master plan which explicitly notes that 'the swamps are good grounds for sugar cane production, and rice and wheat production' (Kapiri Mposhi, District Council, n.d.). Experience of government supported wetland development is also seen in many irrigation and swamp drainage schemes in East and West Africa (see Chapters 2, 4, 7, 9 and 10).

Undoubtedly one of the successes of the global wetland conservation movement has been to raise awareness of the potential risks of wholesale wetland conversion, instilling a precautionary ethic amongst wetland users and planners. However, in this book we argue that wetland conversion, particularly for agriculture, should be one option to be considered on a case-by-case basis given its important contribution to the livelihoods of the rural poor, and the way in which it can stimulate rural development (although we also stress that this must be done within an understanding of the impacts on specific ESS). Roggeri (1995), for example, suggests that the conversion of wetlands may be justified in circumstances where:

- social and economic needs are particularly pressing;
- wetland conversion is the only solution that would meet these needs;
- it has been demonstrated that the wetland could contribute significantly to meet these needs;
- it has been demonstrated that the wetland in its natural state is of minor value (and that its value will not increase in the future).

(Roggeri, 1995: 86)

More recently it has been argued by the MA (2005) that wetland agriculture should be conceptualized as a wetland provisioning service and that careful management that encompasses appropriate water and agricultural practices can result in a net increase in the total ESS/benefits derived from a wetland (McCartney *et al.*, 2005). Such an approach recognizes trade-offs associated with changes in the ecological condition of a wetland (e.g. reduced cultural and regulating services in exchange for increased provisioning services such as food and water) (MA, 2005) and, while accepting the difficulty of identifying and quantifying the full suite of services provided by any given wetland, proposes management approaches that explicitly consider these trade-offs (McCartney *et al.*, 2005).

Overall in the wetland development discourse there has been a gradual transition from an economic development 'at all costs' approach to agriculture in wetlands – focused primarily on raising output – to a realization that agricultural systems are underpinned and depend fundamentally on ecological processes and the services provided by the wetlands. There is now an understanding that environmental and social factors have to be considered in much greater detail in wetland development than was typically the case in the past, in order to ensure environmental sustainability, socially acceptable development and economic sustainability. Nonetheless, the focus here remains on people using wetlands and the contribution of wetlands to development in a sustainable way.

Wetland conservation

The second perspective that has contributed to the wetlands discourse can be described as a conservation/preservation approach. From a conservation biology perspective the goal of wetland conservation is the protection of species and biodiversity, although the underlying rationale for this may be rooted in various ideologies ranging from a recognition of the intrinsic value of biological diversity (a so-called 'deep ecology' approach) to more anthropocentric approaches that value biodiversity from a utilitarian perspective (e.g. the provision of medicines or wildlife tourism) (Pepper, 1996). In theory there are some marked differences between 'conservation' approaches, which tend to be associated with active management and sustainable resource use, and 'preservation', which implies protection from any forms of use. In practice, and in the context of wetland management in Africa, however, the differences between conservation and preservation have often become blurred in recent years (Adams, 2009).

The origins of the modern day wetland conservation movement can be traced back to concerns for waterfowl populations, originating first during the 1930s in North America, and then later during the 1940s in Europe. Recognition of the need to protect wetland habitats for endangered species was central to the establishment of the Wildfowl and Wetlands Trust in 1946, and was also incorporated into the mandate of the IUCN (World Conservation Union, originally known as the International Union for the Conservation of Nature) established in 1948. In 1960, the IUCN received a proposal calling for an international programme specifically dedicated to wetlands, and throughout the decade discussions among various governments and non-governmental organizations (NGOs), particularly the IUCN, the IWRB (International Waterfowl and Wetlands Research Bureau, which later became Wetlands International, WI) and the ICBP (International Council for Bird Preservation, now Birdlife International) explored the wide range of issues affecting wetlands and the feasibility of implementing such a programme (Matthews, 1993). In 1971 this process culminated in the creation of the Ramsar Convention on Wetlands of International Importance especially as Waterfowl Habitat, which set out a process of accession for countries wishing to designate and list wetlands of international importance for conservation purposes (Matthews, 1993; Davis, 1994).

Although space precludes a detailed discussion of the development of the Ramsar Convention (see Matthews, 1993; Farrier and Tucker, 2000; Bowman, 2002), it is important to consider the influence the convention has had on wetland policy-making over the last 40 years, not least because it remains the only international wetland convention and has, until recently, been dominated by the conservation interests of the global north. The conservation focus is evident in its original text that makes it clear that wetlands and their various ecological functions should be protected from human encroachment for the benefit of waterfowl, and that protection should be achieved via the 'conservation of the wetlands included in the list' (Ramsar, 1971: Article 3). Although the text goes on to suggest that countries should also promote 'as far as possible the wise use of wetlands in their territory', thus suggesting that some form of human intervention is acceptable, it would be a further 16 years before the concept of 'wise use' was formally defined and recommended for adoption by contracting parties:

> The wise use of wetlands is their sustainable utilization for the benefit of humankind in a way compatible with the maintenance of the natural properties of the ecosystem.
>
> Sustainable utilization is defined as human use of a wetland so that it may yield the greatest continuous benefit to present generations while maintaining its potential to meet the needs and aspirations of future generations.
>
> Natural properties of the ecosystem are defined as those physical, biological or chemical components, such as soil, water, plants, animals and nutrients, and the interactions between them.
>
> (Ramsar, 1987: Recommendation 3.3)

While in many respects this was a significant step forward in acknowledging the importance of people in wetland utilization (occurring during a period of unprecedented global interest in sustainable development), the wise-use principle remained problematic for several reasons. First, in terms of attempting to embrace sustainable development and bring together conservation and the needs of people, it remained heavily skewed in favour of environmental concerns, not least because of the incompatibility between human use and the maintenance of 'natural properties'. As Farrier and Tucker (2000) point out, 'naturalness' is a state that is difficult to define with regards to wetlands, especially since very few wetlands will have evaded the influence of humans in one way or another. A second related issue is the nebulous nature of natural properties themselves (what Ramsar reinterpreted as 'ecological character' in 1999, and later linked to the MA's 'ecosystem services' in 2005), and the evidence for their occurrence in wetlands.

Finally, another problem with the adoption of the wise-use principle is that it has served to legitimise some forms of human interaction with wetlands but not others, and hence raised issues of equity particularly regarding access to these resources – an issue with particular ramifications for poor people in Africa. Consequently, the forms of acceptable 'wise use' that were being promoted widely in the literature and wetland conservation initiatives during the late 1980s and early 1990s, tended to consist of non-transformational uses of wetlands such as fishing and reed collection, and argued against any intrusion of agriculture into wetlands (Marchand and Udo, 1989; Dugan, 1990). These approaches, often operationalized and branded as Integrated Conservation and Development Projects (ICDPs), actively sought the participation of local people in wetland management strategies, and whilst they were successful in conserving the natural status of many wetlands, their contribution towards livelihoods is unclear. In a review of ICDPs, Sellamuttu *et al.* (2008) suggest a range of problems including poor economic returns from alternative incomes for local people, a poor understanding of the dynamic complexities of people's livelihoods and the local context on the part of conservationists, unrealistic development targets and the lack of monitoring. Overall, there is little evidence to suggest that these initiatives had a major impact on poverty reduction, or addressed other critical needs such as food security (Kellert *et al.*, 2000; Fisher *et al.*, 2005).

One other important, but debatable, achievement of the global wetland conservation discourse has been the promotion of universal wetland values, functions and attributes, the mention of which has become a prerequisite to any government report or academic paper discussing wetlands. The empirical scientific evidence, however, remains sketchy and often contradictory, with some wetlands having the functions widely attributed to them and others not (Bullock and Acreman, 2003; McCartney *et al.*, 2010; McCartney *et al.*, 2011). This has tended to fuel the debate between the conservationists and those concerned with the development of wetlands because of the often unrealistic claims about the ESS present in specific wetlands.

Putting livelihoods into wetland conservation

Since the late 1990s there has been a marked reorientation towards human development in wetland conservation which has been reflected in the themes of the Ramsar Conference of the Contracting Parties (COP) meetings (see Table 1.2) (although prior to COP7, meetings were not assigned a specific theme).

Table 1.2 Recent Ramsar Convention COPs

Year	Location	Theme	
1999	San Jose, Costa Rica	COP7:	'People and Wetlands: the vital link'
2002	Valencia, Spain	COP8:	'Wetlands: water, life, and culture'
2005	Kampala, Uganda	COP9:	'Wetlands and water: supporting life, sustaining livelihoods'
2008	Changwon, Korea	COP10:	'Healthy Wetlands, Healthy People'
2012	Bucharest, Romania	COP11:	'Wetlands: Home and Destination'

Of particular note is the inclusion of 'sustaining livelihoods' in the title of COP9, the first to be held in Africa, and significant for several reasons. First, it integrated the MA's ESS concept into a new definition of wise use: 'Wise use of wetlands is the maintenance of their ecological character, achieved through the implementation of ecosystem approaches, within the context of sustainable development' (Ramsar, 2005: 6).

Second, the meeting was also notable for its attention to river basin planning, a hitherto neglected area of the convention, which critics suggested had resulted in wetlands erroneously being viewed in isolation from upstream and downstream influences. Third, and perhaps most significantly, the Convention recognised the importance of the MDGs as internationally agreed development strategies, and set out actions for contributing to poverty reduction (Resolution IX.14) (Box 1.1). Despite the significant mention of developing 'capacity to use' and 'respecting the rights of local communities', however, the following COP in 2008 acknowledged the lack of clear guidance in this resolution. This was addressed to some degree at that COP by Resolution X.28, which somewhat significantly shifts the emphasis from poverty 'reduction' to poverty 'eradication', and subsequently goes on to emphasise the importance of wetlands for human health, the inclusion of traditional knowledge into wetland management strategies and the potential role that financial incentives could play in wetland management (Box 1.1).

Box 1.1 Wetlands and poverty reduction: Resolutions IX.14 and X.28

a) Detail from Resolution IX.14 (2005) 'Wetlands and Poverty Reduction'

The COP ...

7. Urges all Contracting Parties and other governments to take action to contribute to poverty reduction, especially in the following areas:

- *human life and safety:* measures to protect against impacts such as cyclones, etc. ... through the sustainable use and restoration of wetlands;
- *access to resources:* measures to improve access to and develop capacity to use, on a sustainable basis, land, water and wetland resources ... respecting the rights of local communities and indigenous peoples;
- *ecological sustainability:* measures to enhance the priority given to sustainability in all relevant mainstream policy sectors, including ecosystem restoration measures;
- *governance:* measures to improve the empowerment of the poor;
- *economies:* measures to maintain or improve, on an ecologically sustainable basis, the ecosystem benefits/services that wetlands provide.

Source: http://www.ramsar.org/pdf/res/key_res_ix_14_e.pdf

b) Detail from Resolution X.28 (2008) 'Wetlands and Poverty Eradication'

The COP ...

Urges Contracting Parties, in relation to the framework of actions set out in Resolution IX.14, also to:

- continue to seek to integrate wetland wise use and management, including wetland restoration as appropriate, into all relevant national and regional policies;
- recognize in their planning and land management policies and strategies the role of wetlands in sanitation and human health;
- respect and incorporate traditional knowledge and practices and local perspectives into national wetland management and sustainable livelihood initiatives;
- ensure that any early warning systems and contingency plans established to safeguard people against natural disasters ... include the use of wetland management and, as appropriate, restoration measures;

- collaborate with relevant institutions in developing suitable eco-tourism activities in wetlands;
- collate knowledge on best practices and promote its transfer for the wise use, extraction, processing and marketing of wetland products in order to enhance poverty eradication;
- establish financial incentives or investments such as micro-credit schemes ... that improve wetland management and contribute to tangible poverty eradication;
- encourage the introduction of payments for ecosystem services as a means to raise funds for poverty eradication programmes;
- consider wetland services as economic goods so that their use may be included in tax-based economic mechanisms such as user pays;
- recognize the importance of identifying existing marketing networks and ways to access these; and
- take measures to safeguard peoples' livelihoods derived from wetlands in areas where mining and other extractive industries are taking place.

Source: http://www.ramsar.org/pdf/res/key_res_x_28_e.pdf

This shift in Ramsar thinking has been mirrored in other wetland conservation organizations such as the Netherlands-based WI, which in 2005 established its Wetlands and Poverty Reduction Project (WPRP). Working in three target areas within Africa, the project sought to demonstrate how sustainable wetland management, based on 'wise use', could be integrated into poverty reduction strategies (Wetlands International, 2005). Like the Ramsar Convention, the WPRP highlighted its strategic alignment with the MDGs, which have become increasingly influential in conservation organizations. Indeed, the Millennium Declaration and the universal acceptance of the eight goals and 21 development targets have significantly altered the landscape of international aid, to the extent that it has raised some difficult questions for the conservation movement over whether it is morally justifiable to conserve wetlands in ways that deny people an opportunity to develop their livelihoods in all potential ways (Fisher *et al.*, 2005).

In response, and as suggested above, many wetland conservation organizations have taken steps to reposition their activities to embrace development goals and livelihood outcomes, but concerns have been raised over the extent to which the issues of livelihoods and poverty reduction are being 'tagged on' to conservation programmes as a convenient way of accessing funding. Critically, what even the most enlightened wetland conservation initiatives appear to lack is a 'bottom-up' approach to exploring wetland-based livelihoods, i.e. one which examines the development needs and aspirations of local people, and then seeks to identify how these can be achieved alongside the sustainable use of ESS. One key reason why such an approach is seldom pursued can be summarized in one word: agriculture. Indeed, the transformation of wetlands by agriculture remains a serious concern among most conservationists and one which is well justified as agriculture has long been cited as the greatest threat to wetlands and their conservation or sustainable use (MA, 2005). This is principally because:

- agricultural interventions often have significant impacts on the ecology and hence functioning of wetland ecosystems;
- agriculture, both within wetlands and in their catchments, can cause significant degradation of wetlands (particularly through disruption of hydrology) leading to the loss of many beneficial ESS and ultimately even undermining the agriculture itself;
- agriculture is the leading pressure upon wetlands and, despite often leading to degradation, short-term gains from cultivation to address economic and survival needs are being given precedence over long-term sustainability; and
- increasing development of large-scale agriculture and irrigation (in conjunction with associated infrastructure, e.g. dams) will, without doubt, result in increased pressure on Africa's wetlands in the near future.

The wetland conservation movement, and particularly the Ramsar Convention, has undoubtedly played a significant role in conserving wetlands and some of their associated ESS around the world over the last 40 years. These organizations have also made significant progress in developing and adapting their strategic focus in response to changes in the wider policy-making and development arena. However, in order to seriously address the MDGs, especially poverty reduction and sustainable livelihoods in Africa, it is essential that they face the monumental challenge of agriculture in wetlands and explore how to mainstream agriculture-based livelihoods into sustainable wetland management strategies.

Convergence

Clearly wetland conservation and development perspectives have both given rise to specific challenges for wetland policy and practice. Despite emerging evidence of a recent convergence in thinking, in terms of a focus on human well-being among conservationists and recognition of the multiple dimensions of sustainability among proponents of wetland development, we argue that the challenge remains of developing an approach that integrates environmental and socio-economic development perspectives to ensure sustainable wetland use. It is to the development literature that we now turn to explore other potential contributions to the wetland discourse which may help achieve this goal.

Learning from 'development': linking wetlands to livelihoods

The wetland conservation movement's attempts to integrate livelihoods and development outcomes into wetland management initiatives represent one way of conceptualizing the livelihoods–wetlands nexus. An examination of the discourse of development over the last 30 years, however, reveals some alternatives that place people and livelihoods as the starting point and at the forefront of strategies that seek to balance environment and development outcomes. These approaches, which we review briefly below, have been little considered in the wetland discourse, and it is our view that these could help progress the convergence of thinking towards people-centred sustainable wetland management.

Community-based approaches

Throughout the 1970s many academics and practitioners began to recognise that large-scale development schemes addressing poverty were achieving only mixed results at best, due to a top-down planning process that failed to appreciate or understand the environmental, social and economic complexities of Africa (Chambers, 1990). Consequently, a range of new ideas that included the basic needs approach, participatory bottom-up development and sensitivity to indigenous knowledge (IK) began to be influential in development circles during the late 1970s and early 1980s. Of particular significance was the work of Robert Chambers who asserted that rural poverty often went unperceived because of biased research and fundamental misunderstandings about people and their livelihoods (Chambers, 1983, 1990). Chambers' efforts to promote a 'farmer first' paradigm in development (Chambers *et al.*, 1989) were particularly influential among northern aid agencies and NGOs, and resulted in a shift in focus towards people and local communities themselves in the development process and their interaction with resources and the environment.

A central driver of this community-level approach to development was the interest generated by an emerging (some would say belated) recognition that many local communities drew on a body of IK of their environment to organize NRM activities to support their livelihoods (Howes and Chambers, 1979; Brokensha *et al.*, 1980). Further research throughout the 1980s and 1990s went on to highlight the dynamic and adaptive nature of IK, and, critically, how this underpins adaptive and sustainable, community-based natural resource management (CBNRM)

(Chambers *et al.*, 1989; Tiffen *et al.*, 1994; Warren *et al.*, 1995; Leach and Mearns, 1996; Reij and Waters-Bayer, 2001; Folke *et al.*, 2002). In contrast to neo-Malthusian perspectives that regarded people as degraders of their environment, evidence from the field appeared to suggest that local people could be powerful agents of sustainability and could even rehabilitate their environment (Tiffen *et al.*, 1994). Moreover, this view was also supported by emerging evidence of the important role played by indigenous community organizations in developing local institutional arrangements for managing common pool resources; because they are rooted in IK and community understandings of dynamic local environments and livelihoods, these organizations and institutions are arguably best placed to establish the rules of engagement with natural resources, and in many cases they have been successful in balancing sustainable resource use with livelihood needs (Ostrom, 1990; Blunt and Warren, 1996; Hinchcliffe *et al.*, 1999; Pretty and Ward, 2001; Mazzucato and Niemeijer, 2002). The Institutional Analysis and Development (IAD) Framework developed by Ostrom *et al.* (1994) has also made a key contribution to recent CBNRM approaches by providing a conceptual model for understanding how the interplay of livelihoods, resources and institutional arrangements create patterns of interaction with resources, which themselves lead to specific resource management outcomes (see Koontz, 2003; Sellamuttu *et al.*, 2008).

CBNRM approaches, therefore, have the advantage of placing people, knowledge, institutions and livelihoods as the starting point for environment–development relationships, and in practice they have sought to build capacity from the bottom-up and engender social, economic and environmental sustainability. Nonetheless, they have come under criticism not least because of their perceived variable success in balancing sustainable resource use with poverty reduction (Fisher *et al.*, 2005; Blaikie, 2006). A fundamental issue here, however, relates to the inconsistencies among the goals of these approaches; CBNRM has often been used as an umbrella term for a range of very different project interventions that may be skewed towards either conservation or development outcomes. Yet despite these concerns there are clearly lessons to be learned from CBNRM, and the ideas that have shaped them, for the development of wetland-based livelihoods.

Socio-ecological systems, resilience and sustainable development

Socio-ecological systems (SES) theory has its roots in a systems view of the environment, which emphasises the dynamic and adaptive inter-relationships between biological systems and the physical environment. As such, ecological systems are considered to be both self-organizing and self-maintaining (Berkes and Folke, 1992). When applied to human–environment relationships (or 'human ecology') the natural environment similarly places constraints on human systems, which in turn adapt, organize and modify their relationship with the environment and natural capital in a series of feedbacks. Proponents of an SES approach suggest that these human–environment relationships are site-specific, constantly evolving and fundamentally complex in nature. Critically, however, they tend to behave in 'non-linear', unpredictable ways, which makes management particularly challenging (Folke *et al.*, 2002). In urging the need for a wider appreciation of the complexity of SES, Ostrom (2007) is especially critical of the dominant global discourse of NRM that has tended to pursue simplistic, linear, 'panacea' solutions to resource management problems, rather than seeking 'appropriate types of solutions for specific niches' (Ostrom, 2007: 4). Put plainly, the first key message emerging from the SES literature is that each environment–development scenario is unique, and requires an understanding of the social, economic, ecological and political circumstances that render it as such.

However, as well as arguing the need for more nuanced, diagnostic approaches to environment–development issues (Ostrom, 2007, 2009), advocates of SES approaches have made significant contributions to the discourse of sustainable development in recent years through mainstreaming two important concepts: resilience and adaptive capacity. If sustainable development means sustaining and enhancing the social, economic and ecological benefits derived from SES, then not only do researchers and practitioners need to understand the social, economic and ecological interrelationships that exist, they also need to understand how these complex systems evolve, respond and adapt, particularly given the unpredictability of external events. The capacity of an SES to buffer and absorb shocks and pressures, whilst continuing to function, constitutes its resilience (Adger, 2000; Folke *et al.*, 2002; Berkes *et al.*, 2003). Conversely, where systems have little resilience and hence are considered vulnerable, external pressures can lead to serious and irreversible change, for example in terms of ecosystem degradation or the erosion of social capital. While it is easy to see why, in academic and practitioner circles, 'resilience' and 'vulnerability' have become synonymous with 'sustainable' and 'unsustainable' respectively, the links are not as clear cut; what could be considered a system's resilience to pressure could just as well denote a resistance to change, which is arguably incompatible with the principles of sustainable development. Resilience, rather, infers a more positive dynamic response to change and is central to supporting and enhancing adaptive capacity, i.e. 'the ability of a socio-ecological system to cope with novel situations without losing options for the future' (Folke *et al.*, 2002: 7). In ecological systems, adaptive capacity is represented in terms of biodiversity, where species have colonised and adapted to changes in the physical environment (biodiversity in turn is credited as further enhancing resilience). In social systems, adaptive capacity is manifest in institutional arrangements for common property resources (CPRs), social networks and human capital in terms of knowledge acquisition and innovation (Ostrom, 1990; Berkes, 1999; Adger, 2000; Berkes *et al.*, 2003; Olsson *et al.*, 2004).

The second key contribution of SES to the broader environment–development debate, therefore, is the idea that management interventions should focus on building resilience (both social and ecological) and adaptive capacity as a means of facilitating sustainable development. Folke *et al.* (2002) suggest several priority areas here for policy-makers that include developing indicators of resilience and vulnerability for resource systems, identifying thresholds of change and pressures, and developing adaptive management strategies in which both resource users and policy-makers at different levels 'learn by doing'. However, these require a policy-making and governance environment that is equally adaptive, flexible and capable of simultaneously supporting a range of management options for resource systems.

The Sustainable Livelihoods Approach

The Sustainable Livelihoods Approach (SLA), developed during the 1990s at the Institute of Development Studies (IDS) and the UK Department for International Development (DFID), is a tool for analysing the livelihoods of the rural poor. Drawing particularly on the work of Robert Chambers and Gordon Conway, Scoones (1998: 7) conceptualises a livelihood as

> the capabilities, assets (including both material and social resources) and activities required for a means of living. A livelihood is sustainable when it can cope with and recover from stresses and shocks, maintain or enhance its capabilities and assets, while not undermining the natural resource base.

The Sustainable Livelihoods Framework (SLF) (Figure 1.6) attempts to illustrate how the livelihood strategies pursued by the rural poor are directly related to their local-level assets, but that the use of these assets are influenced by the wider socio-economic, political, cultural and environmental context that include government policies, market forces, institutions and a 'vulnerability context' that includes elements of uncertainty such as environmental or socio-economic change. A sustainable livelihood is one that results in material and financial benefits through increased well-being, improved resilience and enhanced food security but one that critically maintains the environmental sustainability of the resource base.

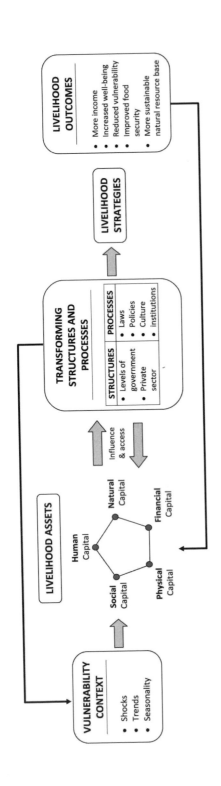

Livelihood assets

Human capital	Skills, knowledge and abilities, health status, availability of labour
Natural capital	Natural resources such as land, forests, fisheries, wetlands, biodiversity
Physical capital	Infrastructure and producer goods; water, energy, transport, shelter, communications
Financial capital	Stocks and flows of money, e.g. savings, remittances
Social capital	Networks and connections with others, membership of civil society groups, relationships of trust and reciprocity

Figure 1.6 The Sustainable Livelihoods Framework

Source: Binns *et al.*, 2011 adapted from DFID, 1999

This framework is, therefore, a useful way of conceptualizing wetland–livelihood relationships principally because it breaks down wetland-based livelihoods into their constituent parts and helps pinpoint the specific factors or processes that are influential in determining the various livelihood outcomes (Ellis, 2000). For example, a sustainable wetland-based livelihood clearly depends upon the availability of natural capital, i.e. wetland ESS, but also potentially social capital, i.e. the relations of trust and reciprocity within a community that may ensure equitable access to wetland resources (see Chapter 4). Similarly, human capital in terms of the labour, knowledge and skills required for wetland use is critical. Where these various assets exist and are strong, the resilience of the livelihood strategy to external shocks and pressures is also increased, and hence the chances of achieving positive livelihood outcomes (including the maintenance of ESS) are maximized.

Despite its enthusiastic reception during the late 1990s, interest in the SLA appears to have waned during the late 2000s with critics questioning its utility beyond the household level, its oversimplification of policy and institutional processes, and the lack of attention paid to intra-household dynamics and gender power relations. Nonetheless, the SLA remains a vital component of sectoral programmes within many large development agencies, and as recently as 2008 was being adopted as a framework for analysis in Wetland International's WPRP (Sellamuttu *et al.*, 2008).

Reflections

What these approaches have in common is an explicit engagement with the environmental, social and economic dimensions of sustainability, and, in addition, a commitment to an equitable and sustainable form of development where the goal is the enhancement of both human and ecological systems. In different ways, they have sought to redress the balance between human well-being and their dependence on natural resources on the one hand, and environmental sustainability on the other, by drawing attention to the complex symbiotic relationships between the two. The emerging lessons here for wetland management in Africa are clear. First, whilst all these approaches emphasise the importance of multi-level stakeholder engagement, the participation of local wetland users in the planning, management and decision-making process is essential; no one understands wetland-based livelihoods better than those who are actually engaged in them, and local users are likely to have developed environmental knowledge and institutional arrangements over time. Second, wetland–livelihood systems are dynamic in space and time, and unpredictable. They are influenced by site-specific ecological feedbacks and local, social and institutional contexts, which the traditional top-down 'one policy fits all' approach would find difficult to accommodate. Again, multi-stakeholder participation in the form of adaptive co-management offers one way forward in decentralizing wetland policy-making. Finally, given the complexity and uncertainty of wetland systems, building resilience at the local level is arguably a fundamental prerequisite to achieving environmentally, socially and economically sustainable wetland livelihoods. Understanding local-level wetland–people relationships and existing institutional arrangements that facilitate socio-ecological resilience should be an entry point for policy-makers.

Recent contributions to wetland thinking

Three studies since 2005 have attempted to move the wetland management discourse forward, albeit with limited direct reference to the development discussion above. These studies, outlined below, have sought in different ways to merge ecological interests with development agendas, and specifically acknowledge the importance of agriculture in wetland livelihoods. Each has sought to integrate these ecological and development aspects into practical management strategies.

The Millennium Ecosystem Assessment (MA)

The Millennium Ecosystem Assessment (MA) in one of its many reports examined wetland functions, services and benefits, and applied to these the new ESS framework (MA, 2005). From a human development perspective it is significant how this framework, emerging from a primarily ecological study, recognises the diverse ways in which ESS contribute to livelihoods and development, especially through wetland provisioning services (Table 1.1). It also identifies how the maintenance of provisioning services is linked to the other three ESS, and how, for instance, undermining supporting and regulating services can reduce the ability of the wetlands to provide provisioning services. Hence, the MA explicitly acknowledges that the ESS of wetlands together will play a key role in achieving the MDGs, and so stresses the need for a holistic view. It recognizes that the development of livelihood benefits from the provisioning services of wetlands can make an important contribution to poverty reduction and that those benefits may be used to justify limited conversion in selected cases. At the same time, it argues that the over-development or over-use of some ESS, notably provisioning services, will lead to the undermining of other ESS (e.g. regulating) and so will make achievement of other MDGs (e.g. safe drinking water and sanitation) problematic. As a result, the MA suggests that balancing ESS in wetlands is essential for their survival.

Whilst this approach has much in common with Ramsar's 'wise use' and 'sustainable utilization' concepts (and indeed the MA itself acknowledges that it provides a 'framework for the delivery' of Ramsar's goals) the approach is arguably more progressive and takes the debate forward in several ways. First, as mentioned above, it emphasizes the importance of human well-being and the ways in which different ESS can support this. Second, it raises the important issue of trade-offs among ESS, and suggests that consideration of trade-offs should be central to the design of interventions that support the MDGs. Third, it advocates the wider use of economic mechanisms, such as the direct economic valuation of wetland services and payment for ESS, arguing that wetlands have tended to be ignored or under-valued by resource users and markets in the past. In spite of these innovations, however, the MA ultimately takes a conservationist approach to the issue of transformational uses of wetlands, especially agriculture, which it regards as a major driver of wetland loss. Rather disappointingly, it also advocates a technocentric approach to management:

> In regions where agricultural expansion continues to be a large threat to wetlands, the development, assessment and diffusion of technologies that could increase the production of food per unit area sustainably, without harmful trade-offs related to excessive consumption of water or use of nutrients or pesticides, would significantly lessen pressure on wetlands.
>
> (MA, 2005: 66)

Although it does goes on to concede that some countries 'lack the financial resources and institutional capabilities to gain and use these technologies', with the exception of occasional reference to 'stakeholder involvement', there is little discussion of the perspective of local people and the role that they and indigenous technologies and institutions can play in sustainable use and management of wetlands.

The Comprehensive Assessment of Water Management in Agriculture (CA)

The Comprehensive Assessment of Water Management in Agriculture (CA) comprised a critical evaluation of the benefits, costs and impacts of the past 50 years of water development, the water management challenges communities face today and the solutions so far developed (CA, 2007). It addressed the overarching question: 'how can water management in agriculture be developed and managed to help end poverty and hunger, ensure environmentally sustainable practices, and find the right balance between food and environmental security?' It was undertaken by the International Water Management Institute (IWMI), in conjunction with a broad alliance of researchers and policy-makers, including, amongst others, the secretariat of the Ramsar Convention, the Consultative Group on International Agricultural Research (CGIAR), the Food and Agricultural Organization of the United Nations (FAO) and the Convention on Biological Diversity (CBD).

The CA considered both the importance of wetlands for agriculture and the adverse impact of agriculture on wetlands, thereby providing a direct link to the MA. In common with the MA, the CA highlighted the fact that the expansion and intensification of agriculture has had many benefits for society, but has also had adverse impacts on ecosystems globally. It also highlighted the fact that agricultural systems depend fundamentally on ecological processes, so if ecosystems become degraded not only are many direct ESS lost but agricultural productivity itself may be undermined (CA, 2007).

In contrast to the MA, the primary focus of the CA was not on maintaining the wetland ecosystems *per se*, but rather on sustainably optimizing the livelihood benefits from those ecosystems. Hence, a key distinction between the CA and the MA is the recognition that managing ecosystems for livelihoods is going to have an impact and this is not necessarily congruent with managing them for biodiversity goals. The CA develops further the idea of conflicts of interest in wetland use and the need for trade-offs between livelihood requirements and conservation that need skilful and innovative forms of management to balance. The objectives of addressing these trade-offs should not be to maximize values for conservation and poverty reduction simultaneously but rather to produce the greatest overall net benefits for people whilst at the same time avoiding fundamental ecological threats and ensuring long-term sustainability of all ESS (Sellamuttu *et al.*, 2008). Thus a pluralistic approach is required that provides opportunities to increase the overall productivity of agricultural systems whilst ensuring that all uses of ESS are enhanced rather than harmed by agricultural development (Nguyen-Khoa *et al.*, 2008).

The CA emphasized that drivers of wetland conversion for agriculture will intensify over the next three decades as populations rise and the demand for increased economic output and food production rises steeply. It also noted that these drivers of change will most likely be exacerbated by climate change and will be most severe in developing countries. In recognition of this the CA stressed the need to identify:

- how the ESS that contribute to agriculture can be enhanced; and
- how agricultural activities can be designed to contribute to ecosystem functioning.

(CA, 2007)

It argued that in future, much greater emphasis must be placed on managing agricultural systems (whether in wetlands or not) as an integrated part of the landscape for multiple rather than single services (i.e. for the full suite of ESS). However, it also recognized that difficult choices will have to be made; win–win situations are rare and better approaches for managing trade-offs are required (CA, 2007).

Guidelines on Agriculture and Wetlands Interactions (GAWI)

The third recent study is an initiative that linked both conservation and development perspectives. This involved the FAO and the Ramsar Convention Secretariat, along with the IWMI, Wetland Action (WA) and WI, exploring how to achieve sustainable use regimes for agriculture in wetlands (Wood and van Halsema, 2008). Between 2005 and 2011, GAWI undertook a number of activities including a meta-analysis of 92 cases of wetland management from around the world.

The GAWI meta-analysis applied the Drivers-Pressures-State Changes-Impacts-Responses (DPSIR) analytical framework (see Smeets and Weterings, 1999) as a tool for analysing the agricultural–wetland interactions in 92 case studies and identified pressure points for change. The study confirmed the growing imbalance in ESS in wetlands as a result of agricultural development. In order to sustain the benefits from wetlands and achieve the rebalancing of ESS as suggested by the MA for sustainable use, it was recommended that all ESS, provisioning and non-provisioning ones, should be put to fruitful use in a wetland or across a wetland network in a stream/river basin taking a landscape approach. It was suggested that this rebalancing of ESS may involve a number of interventions which could include:

- redirecting the drivers of change so that the specific needs of society (which lead to drivers) can be met in other ways – through trade, employment, non-wetland farming development, any of which could reduce the imbalances in ESS and the negative state changes in wetlands or elsewhere in the river basin system;
- diversifying the wetland provisioning services used beyond agriculture – through the addition of fishing, crafts, ecotourism and payment for environmental services (PES), etc., so as to still meet household needs while reducing the pressures from mono-agriculture upon the wetlands and the negative state changes and impacts due to cultivation;
- diversifying the demands on wetlands for different ESS so that non-provisioning services can generate income, especially through PES for regulatory or biodiversity conservation services;
- managing land at the basin level in ways to facilitate the maintenance of a balance of ESS overall, with different ESS provided at different points in the river/stream system;
- improving crop choices to plants that require less alteration of the wetland ESS, e.g. irrigated rice, or taro (*Colocasia esculenta*) as opposed to drainage for cultivation of maize.

From this work it has become clear that across the globe, but particularly in SSA, there is the need for a change in thinking about wetlands. This should involve a move from a situation of competition amongst stakeholders who each seek to achieve mono-ESS use of wetlands to meet their own specific interests – agriculture or biodiversity conservation – to a situation where stakeholders work together to achieve a mix of ESS in wetlands, with mutually advantageous multiple benefits that help ensure the sustainability of all ESS and wetlands in the long term. This will require the setting of priorities and policies for the different ESS to be developed or maintained in different parts of a wetland or along a stream valley system, in order to accommodate the multiple demands made on wetland ecosystems. The DPSIR analysis helps achieve this by mapping out the socio-economic demands for specific ESS and the state changes and impacts that result from the development of these ESS, along with their consequences for the balance of services in a wetland system. Overall a congruent and harmonious functional management strategy must be developed that links each demand from society to a specific ESS but also maintains the balance of ESS overall and ensures long-term sustainability.

Key points from the evolving discourse

To summarise, these three documents show the development of a number of key themes which are important for the development of sustainable wetland management for livelihoods in Africa. These are:

- recognition of the contribution of wetlands to livelihoods and MDGs through provisioning services, especially agriculture;
- identification of the growing threats that agriculture and livelihood development can present to the maintenance of wetlands and their full range of ESS;
- recognition of the linkages amongst ESS and the need to keep a balance or a mix of ESS in order to maintain the functioning of wetlands, i.e. for their ecological sustainability;
- developing values for communities from the different ESS, especially generating values for neglected ESS such as regulatory and biodiversity services, sometimes through PES;
- maximizing total benefits for society from the full range of wetland ESS and not just to maximize the benefits from one alone;
- recognizing the need to explore different trade-offs between ESS in wetlands, in conjunction with the wider landscape, so as to maximize overall benefits and maintain a mix of ESS;
- using trade-off methods to address the competing demands and conflicts that can develop between different interest groups with respect to wetland ESS and so ensure sustainability; and
- recognizing the role of people in the management of wetlands for ESS maintenance as wetland users managing agriculture and other provisioning services so as to maximize and sustain overall benefits.

A way ahead

Building on these key points, especially the MA, there have been several recent contributions to the wetlands discourse that have emphasized the role of wetlands for human benefits, especially for health (Horwitz *et al.*, 2012), for livelihoods and poverty reduction (Kumar *et al.*, 2011) and also for ESS and conservation (Maltby and Acreman, 2011; Finlayson *et al.*, 2011). While all acknowledge the need for increasing recognition of the role of people in wetland management, most remain firmly focused on ecological outcomes. At the same time other recent literature reasserts the central role of people not just as beneficiaries but also as actors in wetland use. These include work from the IWMI (McCartney *et al.*, 2010) which stress the contribution of wetlands to livelihoods, especially agriculture and the challenge of reaching sustainable multiple use in wetlands. Similar works include those by Dixon and Wood (2007) on the role that community-based institutions play in managing wetlands, and by the CPWF (2010) on balancing social welfare and environmental security in wetland-based livelihoods. But while there is clear recognition of people as active users and managers of wetlands, these papers are limited in their identification of specific actions that are needed to achieve sustainable use and development of wetlands for livelihoods and poverty reduction.

In the view of the editors and authors of this book, there is no doubt that people need to be put at the centre of exploring how to progress towards sustainable wetland use and development in Africa, with the balance of ESS maintained to ensure sustainability. In this book a new framework for thinking about wetlands is proposed which sees people as inextricably connected to wetlands. Their inclusion is essential to ensure sustainability with respect to all ESS. We believe that this represents a paradigm shift and is a logical culmination in the evolution of thinking over the past decade. We strongly believe that such a change in thinking is essential if African wetlands are to be managed in a manner that, in the face of upcoming pressures, including climate change and food provision, will safeguard their existence and the essential ESS that they can contribute to livelihoods and development.

This paradigm argues that people must be recognized as active user managers, not as conservation managers and certainly not as top-down government planning managers. Sustainable wetland use will only be achieved when wetland users are empowered and have clear rights as users and managers of wetlands themselves. This will require a supportive policy framework that includes security of access to wetlands, and provides appropriate incentives that make investing in the wetlands worthwhile. Figure 1.7 provides a schematic representation of these ideas.

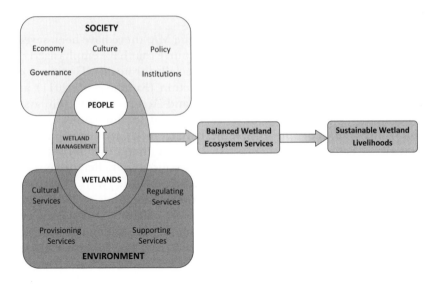

Figure 1.7 A people-centred approach for analysing the management of small
inland wetlands in Africa

The essential elements of a people-centred approach or paradigm include:

- People operating in a framework of society, policies and economic/
  social/cultural incentives.
- People's contribution is reflected in the institutions and coordinated
  management, economic and socio-cultural incentives, as well as policy.
- The critical interaction is how people and wetlands interact. This is
  influenced from both the socio-economic and the environmental
  aspects with their different characteristics.
- The nature of this interaction determines the outcomes in terms of:
  - ecological sustainability – the pattern of ESS;
  - economic sustainability – provisioning services, and others relevant
    to PES;
  - institutional sustainability – arrangements, policies and other
    aspects of socio-institutional organization.

With reference to this framework, and with guidance from the literature
reviewed above, this book seeks to explore the experience with sustainable
wetland management in a number of countries in several parts of Africa. In
these nine case studies the following themes are central:

- The relationship between wetlands, livelihoods and local wetland management arrangements.

- The different dimensions of sustainability (economic, environmental, social and institutional) that need to be integrated in wetland management strategies.
- Technical, institutional and policy elements that can be applied to support sustainable wetland management.
- The need for wetland management strategies across a range of scales in SSA.
- The importance of supportive policy environments for sustainable wetland management.

Chapter 11 of this book reviews the lessons from these case studies and identifies how they may contribute to further elaborating the people-centred paradigm/framework presented above and actions to help achieve sustainable management of inland wetlands in SSA.

References

Abebe, Y. and Geheb, K. (2003) W*etlands of Ethiopia*, IUCN, Nairobi.

Adams, W. M. (2009) *Green Development: Environment and Sustainability in a Developing World*, Routledge, London.

Adekola, O., Morardet, S., de Groot, R. and Grelot, F. (2008) *The Economic and Livelihood Value of Provisioning Services of the Ga-Mampa Wetland, South Africa*, World Water Congress, Montpellier.

Adger, W. N. (2000) 'Social and ecological resilience: are they related?', *Progress in Human Geography*, vol. 24, pp. 347–364.

Andriesse, W., Fresco, L.O., van Duivenbooden, N. and Windmeijer, P. N. (1994) 'Multi-scale characterization of inland valley agro-ecosystems in West Africa', *Netherlands Journal of Agricultural Science*, vol. 42, no. 2, pp. 159–179.

Barbier, E. B. (1993) 'Sustainable use of wetlands – valuing tropical wetland benefits: economic methodologies and applications', *The Geographical Journal*, vol. 159, no. 1, pp. 22–32.

Barbier, E. B., Acreman, M. and Knowler, D. (1997) *Economic Valuation of Wetlands: A Guide for Policy Makers and Planners*, Ramsar Convention Bureau, Gland.

Bell, M. and Roberts, N. (1991) 'The political ecology of dambo soils and water resources in Zimbabwe', *Transactions of the Institute of British Geographers*, vol. 16, pp. 301–318.

Bell, M., Faulkner, R., Hotchkiss, P., Lambert, R., Roberts, N. and Windram, A. (1987) 'The use of dambos in rural development, with reference to Zimbabwe', unpublished report, Loughborough University and University of Zimbabwe.

Berkes, F. (1999) *Sacred Ecology: Traditional Ecological Knowledge and Resource Management*, Taylor and Francis, Philadelphia and London.

Berkes, F., Colding, J. and Folke, C. (eds) (2003) *Navigating Social-ecological Systems: Building Resilience for Complexity and Change*, Cambridge University Press, Cambridge.

Berkes, F. and Folke, C. (1992) 'A systems perspective on the interrelations between natural, human-made and cultural capital', *Ecological Economics*, vol. 5, no. 1, pp. 1–8.

Binns, J. A., Dixon, A. B. and Nel, E. (2011) *Africa: Diversity and Development*, Routledge, London.

Blaikie, P. (2006) 'Is small really beautiful? Community based natural resource management in Malawi and Botswana', *World Development*, vol. 34, no. 11, pp. 1941–1957.

Blunt, P. and Warren, D. M. (1996) *Indigenous Organizations and Development*, ITDG Publishing, London.

Boko, M., Niang, I., Nyong, A., Vogel, C., Githeko, A. and Medany, M. (2007) 'Africa', in M. L. Parry, O. F. Canziani, J. P. Palutikof, P. J. Linden and C. E. van der Hanson (eds) *Climate Change 2007: Impacts, Adaptation, Vulnerability: Contribution of Working Group II to the Fourth Assessment Report of the Intergovernmental Panel on Climate Change*, Cambridge University Press, Cambridge.

Bondestam, L. (1974) 'People and capitalism in the north-east lowlands of Ethiopia', *Journal of Modern African Studies*, vol. 12, pp. 423–439.

Bowman, M. J. (2002) 'The Ramsar Convention on wetlands: has it made a difference?' in O. S. Stokke and O. B. Thommessen (eds) *Yearbook of International Co-operation on Environment and Development 2002/2003*, Earthscan, London.

Brokensha, D., Warren, D. and Werner, O. (1980) *Indigenous Knowledge Systems and Development*, University Press of America, Lanham.

Bullock, A. and Acreman, M. (2003) 'The role of wetlands in the hydrologic cycle', *Hydrology and Earth System Sciences*, vol. 7, pp. 358–389.

CA (2007) *Water for Food, Water for Life: A Comprehensive Assessment of Water Management in Agriculture*, Earthscan and International Water Management Institute, London, Colombo.

Cecchi, A. (1886) *Da Zeila alle frontier del Caffa*, Ermanno Loescher, Rome.

Chambers, R. (1969) *Settlement Schemes in Tropical Africa: A Study of Organisations and Development*, Routledge and Kegan Paul, London.

Chambers, R. (1983) *Rural Development: Putting the Last First*, Longman, London.

Chambers, R. (1990) *Microenvironments Unobserved*, Gatekeeper Series No 22, IIED, London.

Chambers, R., Pacey, A. and Thrupp, L. A. (eds) (1989) *Farmer First: Farmer Innovation and Agricultural Research*, Intermediate Technology Publications, London.

Cook, H. and Williamson, T. (1999) *Water Management in the English Landscape: Field Marsh and Meadow*, Edinburgh University Press, Edinburgh.

CPWF (2010) *Wetlands-Based Livelihoods in the Limpopo Basin: Balancing Social Welfare and Environmental Security*. CPWF Project Report submitted to the Challenge Program on Water and Food (CPWF), http://cgspace.cgiar.org/bitstream/handle/10568/3903/PN30_IWMI_Project%20Report_Mar10_final.pdf?sequence=1 (accessed 2nd August 2012).

Crafter, S. A., Njuguna, S. G. and Howard, G. W. (eds) (1992) *Wetlands of Kenya, Proceedings of the KWWG Seminar on Wetlands of Kenya, Nairobi, Kenya, 3–5th July 1991*, IUCN, Gland.

Daka, A. E. (1997) 'Potentials and limitations of crop production in Dambos (Vleis)', *African Crop Science Conference Proceedings*, vol. 3, pp. 47–55.

Dan-Azumi, J. (2010) 'Agricultural sustainability of fadama farming systems in northern Nigeria: the case of Karshi and Baddeggi', *International Journal of Agricultural Sustainability*, vol. 8, no. 4, pp. 319–330.

Davis, T. J. (1994) *The Ramsar Convention Manual: A Guide to the Convention on Wetlands of International Importance Especially as Waterfowl Habitat*, Ramsar Convention Bureau, Gland.

Denny, P. (1993) 'Wetlands of Africa: introduction', in D. Whigham, D. Dykjova and S. Hejny (eds) *Wetlands of the World: Inventory, Ecology and Management*, vol. 1, Handbook of Vegetation Science, Kluwer Academic Publishers, Dordrecht.

Denny, P. (2001) 'Research, capacity-building and empowerment for sustainable management of African wetland ecosystems', *Hydrobiologia*, vol. 458, no. 1, pp. 21–31.

DFID (1999) *Sustainable Livelihoods Guidance Sheets*, DFID, London, http://www.ennonline.net/resources/667 (accessed 2nd August 2012).

Dixon, A. B. and Wood, A. P. (2003) 'Wetland cultivation and hydrological management in eastern Africa: matching community and hydrological needs through sustainable wetland use', *Natural Resources Forum*, vol. 27, no. 2, pp. 117–129.

Dixon, A. B. and Wood, A. P. (2007) 'Local institutions for wetland management in Ethiopia: sustainability and state intervention', in B. van Koppen, M. Giordano and J. Butterworth (eds) *Community-Based Water Law and Water Resources Management Reform in Developing Countries*, CABI International, Wallingford.

Dugan, P. J. (1990) *Wetland Conservation: A Review of Current Issues and Action*, IUCN, Gland.

Ellis, F. (2000) *Rural Livelihoods and Diversity in Developing Countries*, Oxford University Press, Oxford.

Emerton, L. (2005) *Values and Rewards: Counting and Capturing Ecosystem Water Services for Sustainable Development*, IUCN Water, Nature and Economics Technical Paper no. 1, IUCN, Gland.

FAO (2004) *Wetland Contributions to Livelihoods in Zambia*, FAO – Netherlands Partnership Programme: Sustainable Development and Management of Wetlands, FAO, Rome.

FAO (2006) *Food Security and Agricultural Development in Sub-Saharan Africa: Building a Case for More Public Support*, Policy Assistance Series 2, FAO, Rome.

FAO (2011) *The State of the World's Land and Water Resources for Food and Agriculture (SOLAW): Managing Systems at Risk*, FAO, Rome and Earthscan, London.

Farrier, D. and Tucker, L. (2000) 'Wise use of wetlands under the Ramsar Convention: a challenge for meaningful implementation of international law', *Journal of Environmental Law*, vol. 12, no. 1, pp. 21–42.

Ferreira, R. E. C. (1977) 'Dambos: their agricultural potential III: vegetable and wheat growing', *Farming in Zambia*, vol. 11, no. 1, pp. 30–33.

Finlayson, C. M., Davidson, N., Pritchard, D., Milton, G. R. and MacKay, H. (2011) 'The Ramsar Convention and ecosystem-based approaches to the wise use and sustainable development of wetlands', *Journal of International Wildlife Law and Policy*, vol. 14, pp. 176–198.

Fisher, R. J., Magginnis, S. and Jackson, W. J. (2005) *Poverty and Conservation: Landscapes, People and Power*, IUCN, Gland.

Folke, C., Carpenter, C., Elmqvist, L., *et al.* (2002) *Resilience and Sustainable Development: Building Adaptive Capacity in a World of Transformations*. Scientific Background Paper on Resilience for the process of The World Summit on Sustainable Development on behalf of The Environmental Advisory Council to the Swedish Government. Norsteds tryckeri AB, Stockholm.

Gluckman, M. (1941) *Economy of the Central Barotse Plain*, Rhodes-Livingstone Institute, Lusaka.

Hinchcliffe, F., Thompson, J., Pretty, J., Guit, I. and Shah, P. (1999) *Fertile Ground: The Impacts of Participatory Watershed Management*, ITDG Publishing, London.

Hollis, G. E. (1990) 'Environmental impacts of development on wetlands in arid and semi-arid lands', *Hydrological Sciences Journal*, vol. 35, no. 4, pp. 411–428.

Horwitz, P., Finlayson, C. M. and Weinstein, P. (2012) *Healthy Wetlands, Healthy People: A Review of Wetlands and Human Health Interactions*, Ramsar Technical Report no. 6, Ramsar, Gland and WHO, Geneva.

Howard, G. W., Bakema, R. and Wood A. P. (2009) 'The multiple use of wetlands in Africa', in E. Maltby (ed.) *The Wetlands Handbook*, Blackwell, Oxford.

Howell, P. P. and Allan, J. A. (1994) *The Nile: Sharing a Scarce Resource*, Cambridge University Press, Cambridge.

Howes, M. and Chambers, R. (1979) 'Indigenous technical knowledge: analysis, implications and issues', *IDS Bulletin*, vol. 10, no. 2, pp. 5–11.

Hughes, F. M. R. (1996) 'Wetlands', in W. M. Adams, S. A. Goudie and A. R. Orme (eds) *The Physical Geography of Africa*, Oxford University Press, Oxford.

Inocencio, A., Sally, H. and Merrey, D. J. (2003) *Innovative Approaches to Agricultural Water Use for Improving Food Security in Sub-Saharan Africa*, International Water Management Institute, Colombo.

Kamukala, G. L. and Crafter, S. A. (eds) (1993) *Wetlands of Tanzania, Proceedings of a Seminar on the Wetlands of Tanzania, Morogoro, Tanzania, 27–20th November 1991*, IUCN, Gland.

Kangalawe, R. Y. M. and Liwenga, E. T. (2005) 'Livelihoods in the wetlands of Kilombero Valley in Tanzania: opportunities and challenges to integrated water resource management', *Physics and Chemistry of the Earth*, vol. 30, pp. 968–975.

Kellert, S. R., Mehta, J. N., Ebbin, S. A. and Lichtenfeld, L. L. (2000) 'Community natural resource management: promise, rhetoric, and reality', *Society and Natural Resources*, vol. 13, pp. 705–715.

Koontz, T. M. (2003) *An Introduction to the Institutional Analysis and Development (IAD) Framework for Forest Management Research*. Paper prepared for 'First Nations and Sustainable Forestry: Institutional Conditions for Success' Workshop, University of British Columbia Faculty of Forestry, Vancouver, Canada.

Kumar, R., Horwitz, P., Milton, G. R., Sellamuttu, S. S., Buckton, S. T., Davidson, N. C., Pattnaik, A. K. and Zavagli, M. (2011) 'Assessing wetland ecosystem services and poverty interlinkages: a general framework and case study', *Hydrological Sciences Journal*, vol. 56, no. 8, pp. 1602–1621.

Leach, M. and Mearns, R. (1996) *The Lie of the Land: Challenging Received Wisdom on the African Environment*, James Currey, Oxford.

Lobell, D. B., Burke, M. B., Tebaldi, C., Mastrandrea, M. D., Falcon, W. P. and Naylor, R. L. (2008) 'Prioritizing climate change adaptation needs for food security in 2030', *Science*, vol. 319, pp. 607–610.

MA (Millennium Ecosystem Assessment) (2005) *Ecosystems and Human Well-being: Wetlands and Water Synthesis*, World Resources Institute, Washington, DC.

Makombe, G., Meizen-Dick, R., Davies, S. P. and Sampath, R. K. (2001) 'An evaluation of Bani (dambo) systems as a smallholder irrigation development strategy in Zimbabwe', *Canadian Journal of Agricultural Economics*, vol. 49, pp. 203–216.

Maltby, E. (1986) *Waterlogged Wealth: Why Waste the World's Wet Places?*, Earthscan, London.

Maltby, E. (2009) *Functional Assessment of Wetlands: Towards Evaluation of Ecosystem Services*, CRC Press, Florida.

Maltby, E. and Acreman, M. (2011) 'Ecosystem services of wetlands: pathfinder for a new paradigm', *Hydrological Sciences Journal*, vol. 56, no. 8, pp. 1341–1359.

Maltby, E. and Barker, T. (2009) *The Wetlands Handbook*, Wiley-Blackwell, Chichester.

Maltby, E. and Turner, R. E. (1983) 'Wetlands of the world', *Geographical Magazine*, vol. 55, pp. 12–17.

Marchand, M. and Udo, H. A. (eds) (1989) *The people's role in wetland management*, Proceedings of the International Conference in Leiden, the Netherlands, June 5–8th 1989, Centre for Environmental Studies, Leiden University.

Masiyandima, M., McCartney, M. P. and van Koppen, B. (2004) *Wetland Contributions to Livelihoods in Zambia*, Netherlands Partnership Programme: Sustainable Development and Management of Wetlands, FAO, Rome.

Matthews, G. V. T. (1993) *The Ramsar Convention on Wetlands: Its History and Development*, Ramsar Convention Bureau, Gland.

Mazzucato, V. and Niemeijer, D. (2002) 'Population growth and environment in Africa: local informal institutions, the missing link', *Economic Geography*, vol. 78, pp. 171–193.

McCartney, M. P. and van Koppen, B. (2004) *Wetland Contributions to Livelihoods in United Republic of Tanzania*, Netherlands Partnership Programme: Sustainable Development and Management of Wetlands, FAO, Rome.

McCartney, M. P., Masiyandima, M. and Houghton-Carr, H. A. (2005) *Working Wetlands: Classifying Wetland Potential for Agriculture*, IWMI Research Report 90, IWMI, Colombo.

McCartney, M. P., Morardet, S., Rebelo, L.-M., Masiyandima, M. and Finlayson, C. M. (2011) 'A study of wetland hydrology and ecosystem service provision: GaMampa wetland, South Africa', *Hydrological Sciences Journal*, vol. 56, no. 8, pp. 1452–1466.

McCartney, M. P., Rebelo, L.-M., Sellamuttu, S. S. and de Silva, S. (2010) *Wetlands, Agriculture and Poverty Reduction*, IWMI Research Report 137, IWMI, Colombo.

McFarlane, M. J. and Whitlow, R. (1990) 'Key factors affecting the initiation and progress of gullying in dambos in parts of Zimbabwe and Malawi', *Land Degradation and Rehabilitation*, vol. 2, pp. 215–235.

Melamed, C., Scott, A. and Mitchell, T. (2012) *Separated at Birth, Reunited in Rio? A Roadmap to Bring Environment and Development Back Together*, Overseas Development Institute, London.

Mharapara, I. M. (1994) 'A fundamental approach to dambo utilization', in R. Owen, K. Verbeek, J. Jackson and T. Steenhuis (eds) *Dambo Farming in Zimbabwe: Water Management, Cropping and Soil Potentials for Smallholder Farming in the Wetlands*, University of Zimbabwe Publications, Harare.

Millington, A. C., Helmisch, F. and Rhebergen, G. J. (1985) 'Inland valley swamps and bolis in Sierra Leone: hydrological and pedological considerations for agricultural development', *Zeitschrift für Geomorphologie*, vol. 52, pp. 201–222.

Muller, C. (2009) 'Climate change impact on Sub-Saharan Africa: an overview and analysis of scenarios and models', *Deutsches Institut für Entwicklungspolitik Discussion Paper*, no. 3.

Ndiyoi, M., Sampa, J. B., Thawe, P. and Wood, A. P. (2009) 'Striking a balance: maintaining seasonal dambo wetlands in Malawi and Zambia', in P. Mundy (ed.) *Planting Trees to Eat Fish. Field Experiences in Wetlands and Poverty Reduction*, Wetlands International, Wageningen.

Nguyen-Khoa, S., van Brakel, M. and Beveridge, M. (2008) *Is Water Productivity Relevant in Fisheries and Aquaculture?* Proceedings of the 2nd International Forum on Water and Food, Addis Ababa, 9–14 November, 2008, vol. 2, pp. 6–10.

Olsson, P., Folke, C. and Berkes, F. (2004) 'Adaptive co-management for building social-ecological resilience', *Environmental Management*, vol. 34, no. 1, pp. 75–90.

Osbahr, H., Twyman, C., Thomas, D. S. G. and Adger, W. N. (2008) 'Effective livelihood adaptation to climate change disturbance: scale dimensions of practice in Mozambique', *Geoforum*, vol. 39, no. 6, pp. 1799–2132.

Ostrom, E. (1990) *Governing the Commons: The Evolution of Institutions for Collective Action*, Cambridge University Press, Cambridge.

Ostrom, E. (2007) *Sustainable Social-ecological Systems: An Impossibility?* http://www.indiana.edu/~workshop/publications/materials/conference_papers/W07-2_Ostrom_DLC.pdf (accessed 2nd August 2012).

Ostrom, E. (2009) 'A general framework for analyzing the sustainability of socio-ecological systems', *Science*, vol. 325, no. 5939, pp. 419–422.

Ostrom, E., Gardner, R. and Walker, J. (1994) *Rules, Games, and Common Pool Resources*, University of Michigan Press, Ann Arbor.

Pepper, D. (1996) *Modern Environmentalism: An Introduction*, Routledge, London.

Perara, N. P. (1982) The ecology of wetlands (dambos) of Zambia and their evaluation for agriculture – a model for the management of wetlands in sub-humid eastern and southern Africa', *International Journal of Ecology and Environmental Science*, vol. 8, no. 1, pp. 27–38.

Pretty, J. and Ward, H. (2001) 'Social capital and the environment', *World Development*, vol. 29, no. 2, pp. 209–227.

Ramsar (1971) Convention on Wetlands of International Importance Especially as Waterfowl Habitat. Final text adopted by the International Conference on the Wetlands and Waterfowl at Ramsar, Iran, 2 February 1971, http://www.ramsar.org/cda/en/ramsar-documents-texts-convention-on-20708/main/ramsar/1-31-38%5E20708_4000_0__ (accessed 10th August 2011).

Ramsar (1987) Convention on Wetlands (Ramsar, Iran, 1971). 3rd Meeting of the Conference of the Contracting Parties, Regina, Canada, 27 May–5 June 1987. Recommendation 3.3: Wise use of wetland, http://www.ramsar.org/pdf/rec/key_rec_3.03e.pdf (accessed 10th August 2011).

Ramsar (2005) 9th Meeting of the Conference of the Parties to the Convention on Wetlands (Ramsar, Iran, 1971) Kampala, Uganda, 8–15 November 2005. Resolution IX.1 Annex A: A Conceptual Framework for the wise use of wetlands and the maintenance of their ecological character, http://www.ramsar.org/pdf/res/key_res_ix_01_annexa_e.pdf (accessed 10th August 2011).

Ramsar (2009) 'What are wetlands?' http://www.ramsar.org/cda/en/ramsar-about-faqs-what-are-wetlands/main/ramsar/1-36-37%5E7713_4000_0__ (accessed 10th August 2011).

Rattray, J. M., Cormack, R. M. M. and Staples, R. R. (1953) 'The vlei areas of S. Rhodesia and their uses', *Rhodesian Agricultural Journal*, vol. 50, pp. 465–483.

Rebelo, L.-M., McCartney, M. P. and Finlayson, C. M. (2010) 'Wetlands of sub-Saharan Africa: distribution and contribution of agriculture to livelihoods', *Wetlands Ecology and Management*, vol. 18, pp. 557–572.

Reij, C. and Waters-Bayer, A. (eds) (2001) *Farmer Innovation in Africa: A Source of Inspiration for Agricultural Development*, Earthscan, London.

Roggeri, H. (1995) *Tropical Freshwater Wetlands: A Guide to Current Knowledge and Sustainable Management*, Kluwer Academic Publishers, Dordrecht.

Scoones, I. (1991) 'Wetlands in drylands: key resources for agricultural and pastoral development in Africa', *Ambio*, vol. 20, no. 8, pp. 366–371.

Scoones, I. (1998) *Sustainable Rural Livelihoods: A Framework for Analysis*, IDS Working Paper 72, IDS, Brighton.

Sellamuttu, S. S., de Silva, S., Nguyen Khoa, S. and Samarakoon, J. (2008) *Good Practices and Lessons Learned in Integrating Ecosystem Conservation and Poverty Reduction Objectives in Wetlands*, IWMI, Colombo.

Seyam, I. M., Hoekstra, A. Y. and Ngabirano, H. H. G. (2001) *The Value of Freshwater Wetlands in the Zambezi Basin*, UNESCO-IHE, Institute for Water Education, Delft.

Schuyt, K. D. (2005) 'Economic consequences of wetland degradation for local populations in Africa', *Ecological Economics*, vol. 53, pp. 177–190.

Smeets, E. and Weterings, R. (1999) *Environmental Indicators: Typology and Overview*, European Environment Agency Technical Report no. 25, European Environment Agency, Copenhagen.

Temper, L. (2010) 'Let them eat sugar: life and livelihood in Kenya's Tana Delta', in CEECEC (eds) *CEECEC Handbook: Ecological Economics from the Bottom Up*, http:// www.ceecec.net/wp-content/uploads/2010/11/THE-CEECEC-HANDBOOK. pdf (accessed 2nd August 2012).

Tiffen, M., Mortimore, M. and Gichuki, F. (1994) *More People, Less Erosion: Environmental Recovery in Kenya*, John Wiley, Chichester.

Trapnell, C. G. and Clothier, J. N. (1937) *The Soils, Vegetation and Agriculture Systems of North-Western Rhodesia*, Government of Northern Rhodesia, Lusaka.

Turner, B. (1986) 'The importance of dambos in African agriculture', *Land Use Policy*, vol. 3, no. 4, pp. 343–347.

Turpie, J. K., Smith, B., Emerton, L. and Barnes, J. (1999) *Economic Value of the Zambezi Basin Wetlands*, Report to IUCN ROSA, Harare.

UNDP (2012a) *MDG Report 2012: Assessing Progress in Africa toward the Millennium Development Goals*, http://web.undp.org/africa/documents/mdg/2012.pdf (accessed 11th November 2012).

UNDP (2012b) *Food Production and Consumption Trends in Sub-Saharan Africa: Prospects for the Transformation of the Agricultural Sector*, UNDP, New York.

UNEP (2006) *Africa Environment Outlook 2: Our Environment, Our Wealth*, UNEP, Nairobi.

UNICEF (2006) *Meeting the MDG Drinking Water and Sanitation Target: The Urban and Rural Challenge of the Decade*, UNICEF/WHO, Geneva.

Verhoeven, J. T. A. and Setter, T. L. (2010) 'Agricultural use of wetlands: opportunities and limitations', *Annals of Botany*, vol. 105, no. 1, pp. 155–163.

Warren, D. M., Slikkerveer, L. J. and Brokensha, D. (1995) *The Cultural Dimension of Development: Indigenous Knowledge Systems*, ITDG Publishing, London.

Wetlands International (2005) *The Wetlands and Poverty Reduction Project: Linking Wetland Conservation and Poverty Alleviation*, Wetlands International, Wageningen.

Whitlow, R. (1990) 'Conservation status of wetlands in Zimbabwe: past and present', *GeoJournal*, vol. 20, no. 3, pp. 191–202.

Wood, A. P. and van Halsema, G. E. (eds) (2008) *Scoping Agriculture-Wetland Interactions: Towards a Sustainable Multiple Response Strategy*, FAO, Rome.

Wood, A. P., Abbot, P. G., Hailu, A. and Dixon, A. B. (2001) 'Sustainable management of wetlands: local knowledge versus government policy' in M. Gawler (ed.) *Strategies for the Wise Use of Wetlands: Best Practices for Participatory Management,* Wetlands International, Wageningen.

Woodhouse, P., Bernstein, H. and Hulme, D. (2000) *African enclosures? The Social Dynamics of Wetlands in Drylands,* James Currey, Oxford.

2 The value of wetlands for livelihood support in Tanzania and Zambia

Matthew McCartney

Summary

This chapter presents findings from an investigation of wetland use conducted in four wetlands: two in Tanzania and two in Zambia. The study results were disaggregated using 'wealth' as a simple indicator of socio-economic status. The study results highlight: 1) the reliance of communities on wetland agriculture and natural resources, for both direct cash income and contributions to food security; 2) the significant contribution wetlands make to coping strategies during times of food scarcity; 3) variations in patterns of use, which are the result of subtle differences in environmental and socio-economic conditions; 4) significant changes in patterns of use in recent years which have, in places, been driven by immigration to wetlands, arising in part because of the perceived agricultural opportunities that they provide; and 5) that communities identify a wide range of biophysical and socio-economic constraints to the use of wetlands for agriculture but negative environmental impacts are not generally amongst them. In addition, the nature of household dependence and the demands on wetland resources were found, not unexpectedly, to vary between different socio-economic groups: better-off households accrue different benefits to the poorest. Sustainable development requires policies that empower local people to manage and control the wetlands in their own landscape. Future policies must promote self-regulating and self-enforcing incentives for sustainable management that ensure equitable resource use and safeguard key ecosystem services (ESS).

Introduction

Wetland resources are crucial to the livelihoods and well-being of millions of people throughout sub-Saharan Africa (SSA) (Woodhouse *et al.*, 2000; MA, 2005; Zwarts *et al.*, 2005; Bikangaga *et al.*, 2007; Rebelo *et al.*, 2009). Wetlands and their surrounding catchments support rural living through provision of a range of ESS: water, food, medicines, construction materials and fuel. In addition, wetlands are often a valuable agricultural resource and many cropping and pastoral systems depend on wetlands to varying degrees. As populations grow, uplands become ever more degraded and as the effects of climate change are increasingly realized, wetlands are becoming more important for agriculture (McCartney *et al.*, 2010).

In these circumstances, it is important to know the extent to which patterns of use and the benefits derived from wetlands are changing and to determine how the ESS that they provide can be optimized to ensure long-term environmental and social sustainability as well as the equitable distribution of benefits. One of the major constraints to the sustainable use of African wetlands is lack of knowledge by government planners, natural resource managers and local communities themselves, of the diverse benefits that they provide and techniques by which they can be utilized in a sustainable manner. There is a need for more information on both the monetary and non-monetary value of the ESS provided by wetlands, and the factors that influence peoples' access to and control over wetland resources.

There is significant scope for increasing agricultural production in the wetlands of SSA (Owen *et al.*, 1995; FAO, 1998; McCartney *et al.*, 2010; Kiepe and Rodenburg, this volume). To do this sustainably and equitably without causing degradation and undermining other ESS requires the right mix of policies and incentives. Policies are needed that: 1) address the context and the changing conditions that are encouraging increased use of wetlands for agriculture; and 2) improve livelihoods and simultaneously promote the wise management of land, water and other wetland resources. Key is convincing farmers that in the long term they will be better-off if they protect strategic wetland features, even if this means foregoing immediate short-term benefits. This is not easy when people's survival is insecure and constrained by limited access to resources. Current policy formulation is hampered by a lack of understanding of wetland dynamics, including the role people play in changes occurring and, as yet, a limited understanding of what form the policies should take (Ramsar, 2007).

The study reported here was conducted as part of an investigation of the diverse contributions that wetlands make to the livelihoods of rural people in both Tanzania and Zambia. This chapter presents a synthesis of the information and data gathered in two case studies undertaken in both countries in 2003. The study contributes to the growing body of evidence that illustrates the important contribution that wetlands, and in particular wetland agriculture, make to livelihoods in southern Africa (Bell *et al.*, 1987; Scoones, 1991; Owen *et al.*, 1995; Scoones *et al.*, 1996; Schuyt and Brander, 2004; Emerton, 2005; Adekola *et al.*, 2008; McCartney *et al.*, 2011a). In addition, the study findings have important implications for policy formulation in both countries.

National context

In both Tanzania and Zambia freshwater wetlands are an important landscape feature, covering approximately 7 per cent (79,450 km^2) and 20 per cent (150,520 km^2) of each country respectively (Bakobi, 1993; Mukanda, 1998). In both countries poverty is largely a rural phenomenon and many communities are extremely dependent on natural resources. Although exact numbers are unknown, wetland products and services are very important to the livelihoods of many people (Kamukala and Crafter, 1993; Mukanda, 1998). Since the countries gained independence in the 1960s, various policies and political factors have influenced natural resource utilization and wetlands.

Both countries have ratified the Convention on Wetlands of International Importance, called the Ramsar Convention: Zambia in December 1991 and Tanzania in August 2000. This means that ostensibly both countries are committed to sustainable wetland management, underpinned by a core guiding principle of 'wise use'. This is promoted by the Convention as a pragmatic strategy for promoting the conservation and sustainable use of wetlands (Matthews, 1993). However, despite broad guidance published by the Convention, how to optimize the sustainable benefits that can be achieved, within the ever-changing social, economic and ecological context of any particular wetland, remains far from clear. This is particularly the case in developing countries where poverty reduction is a priority and immediate development needs are acute. Furthermore, countries such as Tanzania and Zambia lack the financial and human resources required to conduct the in-depth studies needed for the development of detailed wetland management plans.

In common with much of SSA, factors such as structural adjustment programmes and economic liberalization in the 1990s have affected the rural productive and marketing infrastructure and have had a significant impact on both agrarian and non-agrarian livelihood strategies in both countries (Bryceson, 1998). The economies of both Tanzania and Zambia depend heavily on agriculture and in each country farming accounts for over 80 per cent of employment.

In Tanzania, a unique policy that had a key influence on wetlands, and other natural resources, was the Arusha Declaration of 1967. This made the village central to development planning and by the early 1970s rural populations were increasingly aggregated into planned settlements, so called Ujamaa Villages. A critical feature of 'villagization' was the registration of villages as corporate entities under the Villages and Ujamaa Act, 1975. This policy was intended both to facilitate provision of water, education and health facilities, and to make way for egalitarian landholding and cooperative farming in accordance with President Nyerere's vision of raising agricultural productivity through gradual collectivization (Campbell, 1996). The concentration of populations had significant implications for natural resource exploitation and in many places had long-term negative environmental impacts (Kikula, 1997; Kangalawe, 2001; Liwenga, 2003). It also had a significant impact on institutional arrangements with authority and governance largely vested within a legalized institutional framework, comprising a Village Council of elected officials. In 1982, with the reintroduction of District Councils, Village Councils became a statutory arm of local government. In the process the remaining pre-colonial or colonial traditional authority (TA) was largely undermined (Lwoga, 1985). In recent years, although the original tenets of villagization have greatly diminished, both through design and in default of assistance from the national government, the institutions of 'village' and village government have consolidated. This has occurred in part because, over the last 20–30 years, many community assets (e.g. schools, clinics and wells) have been communally established and maintained. Consequently, social relations and the community identity are now largely centred on the village and the Village Council. In addition to changes in governance, land tenure has also changed in recent decades from a system of communal property to state ownership (so called 'village land'). Currently individuals and groups are able to make use of village land resources, but only through the forbearance of the Village Council (Campbell, 1996).

More recently, the National Lands Policy (1995) classifies wetlands as 'hazardous land' (i.e. areas that need to be protected because development is likely to pose a danger to life or lead to environmental degradation). The Agriculture and Livestock Policy (1997) recognizes that environmental issues cut across sectors and that agriculture depends on the natural resource base. It recommends a coordinated approach to the utilisation of natural resources. The National Environment Policy (1997) aims to ensure that environmental resources are managed well, including improved management and conservation of wetlands. The policy identifies some problems related to wetland management including land degradation and pollution of water.

In Zambia, a national wetland strategy and action plan has been developed by the Environmental Council of Zambia. This acknowledges both the inherent ecological importance of wetlands as well as the socio-economic value of a range of ESS that they provide (Government of the Republic of Zambia, undated). To support the implementation of the strategy a draft National Wetland Policy was developed in 2001, but to date this has not been formally ratified. Nevertheless, wetlands in Zambia are in principle managed by the state though formal government institutions such as the Zambia Wildlife Authority (ZAWA) and the Fisheries Department. However, in contrast to Tanzania, traditional authorities (lineage chiefs) retain considerable authority and administer their areas of jurisdiction through village headmen. They are empowered to allocate land and resolve disputes over natural resource use. For local people they represent ultimate authority and they are, by far, the most important institutions in the practical management of many wetlands (McCartney et al., 2011b). The recent National Policy on Environment

(Government of the Republic of Zambia, 2009) recognizes the need to safeguard the ecological, economic and social values of wetlands and, significantly, recognizes the importance of involving local communities in the management of natural resources.

In both countries community-based natural resource management (CBNRM) approaches have become increasingly common in recent years and, in theory, provide the opportunity for communities to become actively involved in developing and implementing plans to manage their own resources. In Zambia, this trend of increasing participation from local actors is reflected in the draft national strategy on Environment and Natural Resources (2008–2010), which identifies the conservation and sustainable use of biodiversity in wetland ecosystems as a focal area and the creation of local community institutions to regulate access and use of wetland-based resources as a key activity (Government of the Republic of Zambia, 2008).

There is a large discrepancy between government policies and what actually happens 'on the ground' in both countries. It is evident that neither country has a coordinated inter-sectoral approach to wetland management and use, and in both Tanzania and Zambia policies that are intended to have a direct impact on wetlands have generally emphasized their protection, rather than utilization. There are also recognized shortfalls with current CBNRM strategies, not least of which is failure to adequately involve local communities. A recent study in Zambia concluded that there was only superficial devolution of decision making to communities, amounting in practice to little more than cooption into decisions already taken by government agencies (Nkhata and Breen, 2010). Although it is generally agreed that across both countries wetland resources are currently undergoing significant modification, paucity of data means that the extent of change is unknown.

Description of the case studies

The wetlands chosen for the research (Figure 2.1) covered a range of wetland types and varied considerably in their size, physical characteristics, climate and ecology. In each case study the research was undertaken in a number of villages in which local people relied on the wetland (Table 2.1). Although the importance of other types of ESS was acknowledged, the focus of the study was primarily on provisioning services. For each case study, information was obtained from a range of participatory techniques in combination with a detailed household survey. Broadly speaking, the participatory assessments were intended to provide qualitative insights into the wetland as a whole, while the household surveys were intended to generate more standardized information to highlight patterns about individual and household uses, and to triangulate findings from the participatory assessments.

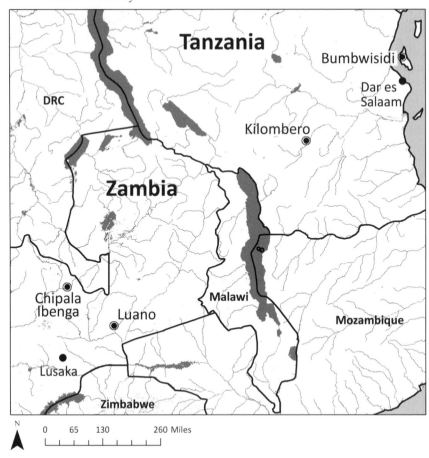

Figure 2.1 Location of the wetland study areas

In order to capture intra-community differences, all fieldwork started with participatory wealth ranking. This involved community members identifying criteria (not only income, but also assets and size of landholdings, etc.) for distinguishing three wealth classes: 'better-off', 'intermediate' and 'poor'. The wealth status of all respondents, whether participating in focus group discussions or in the household surveys, was determined using these criteria. The participatory assessments (e.g. focus group discussions, participatory mapping, transect walks, preference ranking, key informant interviews, etc.) were based on techniques of Participatory Rural Appraisal (PRA) as described in the SEGA Field Handbook (FAO, 2001).

The household surveys were conducted using a structured questionnaire, which was designed to gather specific information on a range of socio-economic factors pertaining to constraints and opportunities within

Table 2.1 Summary of characteristics of the case study wetlands

Name	Location	Wetland area (km²)	Mean annual rainfall (mm)	Villages participating	Description
Zambia					
Chipala-Ibenga	13°14' S 28°27' E	1.0	1,294	Chibilikita	Located in the Copperbelt Province in Masaiti district, agricultural activities in the wetland focus primarily on the cultivation of horticultural crops (leafy vegetables and tomatoes). There are two markets in the vicinity of the wetland, located about 5 km and 32 km away. Externally driven interventions have influenced the use of the wetland. The Cooperative League of the USA (CLUSA) is supporting farmers by introducing treadle pumps for irrigation. The introduction of maize seeds and cassava multiplication by the Program Against Malnutrition impacts the wetland because some farmers grow these crops there.
Luano Valley	14° 26' S 29°58' E	40.0	890	Mwenda, Mikwa, Lukasashi, Mufumbe, Nduani, Ketetaula, Chityoka and Chikanga	Located in the Mkushi district in central Zambia the wetland is situated in a Game Management Area (GMA). As a result, extension support services were very limited. The only public organization active in the area was the Luano Valley Development Program (LVDP) undertaken by a small local non-governmental organization (NGO). The Catholic Mission was the main organization that assisted farmers.

Continued

Table 2.1 Summary of characteristics of the case study wetlands, *continued*

Name	Location	Wetland area (km²)	Mean annual rainfall (mm)	Villages participating	Description
Tanzania					
Bumbwisidi	06°03′S 39°17′E (i.e. on Unguja Island, Zanzibar)	5.6	1,833	Bumbwisudi, Kizimbani, Mwakaje, Miwami, Mkanyangeni and Ndagaa	Seven villages are located close to the wetland. All are traditional (i.e. not Ujamaa) villages. The population in the area rose from 7,232 in 1988 to 11,973 in 2002. About 390 ha of the wetland was used for paddy cultivation. In the early 1980s, 17 boreholes were drilled in the wetland to supply water for 150 ha of rice irrigation. The boreholes also supplied household water to some of the villages. However, only four of the boreholes were still operational in 2003 and only 30 ha of rice were irrigated. The wetland was also used for cultivating sweet potatoes, cassava and vegetables as well as cattle grazing and some fishing.
Kilombero Valley	08°40′S 36°10′E	13,520	1,300	Idete and Siganli	Traditionally the people in these villages were fishermen who had small, cultivated plots in the drylands. Since Tanzanian independence, there has been major development in Kilombero District including sugar cane farming and the construction of a hydropower plant. This resulted in the expansion of the economic sector and, because of the availability of cultivable land, there has been significant immigration to the valley. The population of Kilombero District increased from 71,826 in 1967 to 322,779 in 2002 and both Idete and Siganli experienced significant population increase. The creation in 1992 of the Udzungwa Mountain National Park, to the north of the villages, increased pressure on the wetland because villagers' access to forest resources were curtailed. Activities within the wetland itself are constrained by the presence to the south and east of the Kilombero Game Control Area and the Selous Game Sanctuary.

households for the utilisation of the wetlands. For the household surveys, it was left to individual study teams to decide whether purposive sampling (i.e. selecting randomly a more or less equal number of households *within* each wealth class) or random sampling and determining the wealth class during the interviews, was most appropriate. The former requires that a full list of all community members and their wealth status is easily available. The latter does not require this detailed information, but entails the risk that the number of respondents within a particular class is low.

Contributions to livelihoods/well-being

The studies revealed that, as might be expected, the wetlands were utilized in very different ways and the contribution of wetland agriculture to livelihoods varied significantly from place to place.

Provisioning services

In each case there were differences in the way the wetlands were utilized between the different socio-economic groups (Figure 2.2). For some wetland uses there was a clear stratification on the basis of wealth. For example, with the exception of Bumbwisudi, where even the poorest households own cattle, the use of the wetland for livestock grazing was restricted to those households wealthy enough to purchase and keep livestock. In contrast with other uses (e.g. use of natural vegetation) there was less differentiation between wealth classes, indicating that factors other than wealth (e.g. distance of the household to the wetland, availability of labour or need for these services) were more important determining factors.

With the exception of Chipala Ibenga, where the better off did not use it, at all locations all wealth classes used the wetlands as a source of domestic/drinking water. In all cases there were alternative water sources (i.e. boreholes or hand-dug wells) but villagers stated that the wetlands were a more reliable source of water in the dry season and during droughts.

In all the wetlands a variety of vegetation types were used for a range of different purposes. In the Luano Valley, trees growing in the wetland (local names: *mopani* and *munsamba*) were used for fuel and construction poles. Other trees (local names: *mutototo* and *muswende*) were used for medicine and a variety of grasses were used for thatching. At Chipala Ibenga, some plants were used for medicine (local names: *munsango* and *ntongolya*) and sedges and grasses were used for thatching houses and for making mats and handicrafts (e.g. baskets and hats). Trees were used for fuel wood and wild fruits (local names: *mpundu* and *masuku*) were collected both for local consumption and for sale. In addition, the tubers of *chikanda* (a wild orchid, which is eaten as a delicacy in Zambia) were collected and sold at local markets. In the Kilombero Valley, a number of wetland grasses (local names: *mbasa*, *chekele*, *lusano* and *mibugubu*) were used for thatching and a

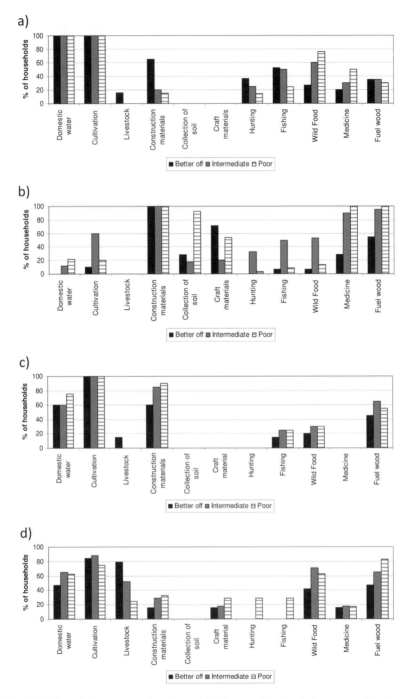

Figure 2.2 Use of different wetland provisioning services by different wealth classes for a) Luano, b) Chipala Ibenga, c) Kilombero and d) Bumbwisudi

large number of timber and other species (local names: *migolegole, mipingo, miwanga, misolwa, misekese, miamata, ming'eng'e, mipulu, mitogo, miombo, milama, mikolekole, mikuyu, mitalo, mizaizai* and *milala*) were used for construction materials. There was also some collection of wild fruits and wood for fuel. From the Bumbwisudi wetland, tree species (local names: *mkarati, mnazi, mdamdam, mpera, mgulabi, mzambarau* and *mikarafuu*) were used for construction materials (i.e. for poles, furniture and thatching) and fuel wood. A range of wild fruits (local names: *miembe, mifuu, minazi, mizambarau, mipera, mitufaa* and *mipapai*) and wild vegetables (local names: *mtoriro, kikwayakwaya, kilembe cha bwana, mwage* and *mrenda*) were also collected.

Fishing was undertaken to varying extents in all the wetlands and, with the exception of Bumbwisudi where it was restricted to the poor, fishing was undertaken by all wealth classes. In all cases the bulk of fish caught were consumed by the household but a small fraction were sold. In the Luano Valley, fish caught include bottle fish, tiger fish and bubble fish. At Chipala Ibenga, worms used for bait were collected from the wetland. In the Kilombero Valley, fishing was a traditional activity of people living in the area. In recent years, population growth and migration of people to the area have led to an increase in the demand for fish. Relatively recent restrictions in hunting are also believed to have led to increased fishing. The study showed that approximately 15 per cent of better-off households and 25 per cent of both intermediate and poor households participated in fishing activities. In the Bumbwisudi wetland, fishing activities were limited to a small number of poor households who caught catfish from ponds and canals as well as other waterways. This was primarily for subsistence and very little was sold.

Contribution to household income

With the exception of Chipala Ibenga, where use of the wetland to collect construction materials dominates, and which is very small so that only a few households can utilize it for agriculture, cultivation is a major activity in all the wetlands. Comparison of cash income generated from the wetlands with other income sources illustrated considerable differences between the case studies in terms of both absolute income, and the proportion, generated through wetland activities (Figure 2.3).

In the Luano Valley, the overall contribution of the wetland to total income was 37 per cent of the approximately US$495 per household per year. However, this average masks important differences between wealth classes (Figure 2.3a). For poor households, 28 per cent of their income was generated from the wetland. In contrast the intermediate and better-off households obtained 43 per cent and 36 per cent of their total income respectively from the wetland. At Chipala Ibenga, total annual incomes were significantly lower across all wealth classes, but the proportion generated from the wetland was much greater: 90 per cent, 68 per cent and 73 per cent

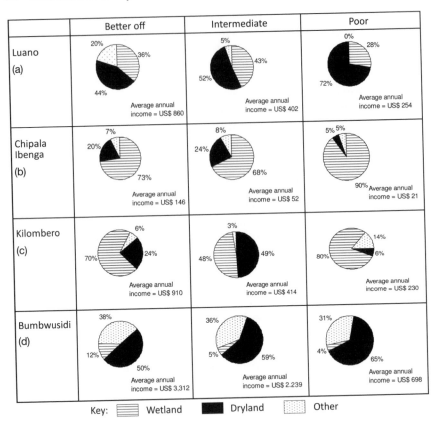

Figure 2.3 Income generated from wetland, dryland and other sources for the four wetlands

for poor, intermediate and better-off households respectively (Figure 2.3b). For all wealth classes the contribution of income generated either through dryland cultivation or from other sources (e.g. small-scale trading, casual labour and remittances) was very small. This illustrates the vital financial contribution – often, but not always, derived through agricultural activities – that wetlands can make to the livelihoods of the very poorest.

Although the communities were financially slightly better off than those in Zambia, similar differences were seen in the Tanzanian case studies. Incomes in Kilombero were most similar to those of the Luano Valley. However, in this case, all wealth classes were much more dependent on the wetland, with better-off, intermediate and poor households generating 70 per cent, 48 per cent and 80 per cent of their income respectively from the wetland (Figure 2.3c). The households of the Bumbwusidi wetland were the wealthiest of all the case studies. However, in this case, the contribution of the wetland to income was relatively small (Figure 2.3d). For poor households only 4 per cent of their income was generated from the wetland. In contrast, the intermediate and better-off households obtained 5 per cent and 12 per cent of their total income respectively from the wetland. The relatively low contribution of the wetland to income, across all the wealth classes, resulted from the fact that the wetland was used primarily for growing rice, the staple food. Generally, only the wealthier households had wetland plots large enough to grow surpluses for sale. For all wealth classes, the main commercial crops (cassava and bananas) were grown primarily on the dryland. Furthermore, in all wealth classes, a large proportion of total income was generated from non-wetland activities (employment, small-scale trading, remittances and, for some of the better-off households, non-wetland related tourism).

The differences between the case studies, in relation to both absolute income and the relative contribution from the wetlands, are explained by variation in both biophysical conditions and socio-economic opportunities and constraints. For Bumbwisidi, the relatively high rainfall provides opportunities for dryland cultivation and greater diversification of crops. Furthermore, the easily accessible markets (there are good roads to Zanzibar town, just 15 km from the wetland) ensure that produce can be sold relatively easily. In contrast, in the other case studies, farming opportunities were hampered, to varying degrees, by lower rainfall (which limits what can be grown), poor communications and lack of market opportunities. In the Kilombero study, people complained that protected land in the vicinity of the villages, which are located close to the Udzungwa Mountain National Park, the Kilombero Game Control Area and the Selous Game Sanctuary, limited their access to forest resources and restricted opportunities outside the wetland.

Contribution to food security

In each case study, the contribution the wetland makes to food security was investigated. In the Luano Valley, where dryland farming is often affected by drought and there are few opportunities to purchase food, all households in all wealth classes reported that at least some of their food was derived from wetland agriculture and ranked wetland plots as their most important food source. At Chipala Ibenga, dryland maize was the staple food and the wetland was used primarily to grow vegetables. Although a proportion of these were sold by all wealth classes, it was found that the poor retained a much greater proportion for home consumption than either intermediate or better-off households (55 per cent, 15 per cent and 12 per cent respectively). In the Kilombero study, although a breakdown by wealth class was not possible, it was found that the wetland contributes a proportion of the food intake of 98 per cent of all households, regardless of wealth class. This compares with 50 per cent of households that obtain some of their food from the dryland and by purchasing. In the Bumbwisudi study, it was found that the wetland provided between 31 per cent and 40 per cent of food for all wealth classes, with other sources being the dryland, purchasing and provision by relatives.

In the Tanzanian studies the extent to which each of the wetlands was used as a coping strategy during periods of food shortage was also investigated. At Bumbwisudi, 58 per cent, 59 per cent and 67 per cent of better-off, intermediate and poor households respectively stated that the wetland was used to cope with food shortages. Similar results were obtained from the Kilombero study. Here, 50 per cent, 65 per cent and 45 per cent of better-off, intermediate and poor households respectively stated that the wetland was used to cope directly with food shortages. Other coping strategies identified in the Kilombero study were through business and crop diversification for the better-off and intermediate households, and through casual labour and reducing the daily number of meals for poor households.

Management and access to wetlands

Given the importance of the benefits derived from wetlands, it would be surprising not to find endemic management practices and forms of social control amongst the wetland-using communities. In the past, indigenous management practices depended on the ability of communities to make and defend management rules. This required effective and credible local authorities; typically traditional leaders who derived authority from their ancestors. In recent decades, in common with the rest of Africa, both Tanzania and Zambia have experienced a radical socio-political transformation (Campbell, 1996; Kikula, 1997; Kangalawe, 2001; Kokwe and Mulolani, 2003; Liwenga, 2003). Now, in both countries authority is, in theory, vested in a range of formal government structures and wetland utilization is nominally regulated by a wide range of national policy and legislation (see above).

In Zambia, village heads and/or chiefs officially remain custodians of the land on behalf of the state. The chiefs continue to grant rights to occupy and use land, impose restrictions on the use of certain areas (e.g. prohibiting cultivation or the grazing of animals in places) and resolve disputes. Under the current Land Act (1993), allocation by a chief can be formalized by obtaining title deeds (with a lease of up to 99 years if a rigorous boundary survey has been conducted and 14 years if only a sketch plan is available) but this can be a long, drawn-out process and in reality statutory title may be no more secure than an unrecorded customary grant. Whether or not the land is held by customary grant or statutory title it can be inherited by next of kin (Adams, 2003).

At both Chipala Ibenga and Luano, local chiefs allocate wetland plots for cultivation. It is not clear how many households had obtained title deeds, but in both wetlands it is probably very few. In both cases, all community members had free access to water for domestic use and also to natural vegetation for food, fuel and medicine. In theory permission to hunt or fish requires that a permit is obtained from the ZAWA. In practice, in both cases, hunting and fishing is done largely illegally, but almost certainly with tacit approval of the chiefs.

In both Tanzanian case studies, the Village Council was identified as the paramount authority on wetland utilization. In the Kilombero Valley, the Village Council specifies who can cultivate and where. Local byelaws stipulate how much land individual households can cultivate. Access to land is lost if the land is not cultivated for two years in succession and the Village Council may then re-allocate the plots. In Kilombero, after harvesting, cattle can graze in areas that were previously cultivated. In other months, to protect crops, the councils have identified wetland areas that are set aside for grazing. In one village the council had identified an area of woodland within the wetland that was protected to provide fuel wood. In the other wetland sites no specific area was identified, but certain trees were protected by local byelaws. For fishing and hunting activities, permits are required from the regional authorities, though again it is not clear to what extent these are actually obtained.

Of the four case studies, the Bumbwisudi study was unique because although the Village Councils were identified as the principal institution controlling resource use within each village, there was also an informal farmers association, the Bumbwisudi Irrigation Group (BIG), which, to a large extent, managed cultivation within the wetland. Established in 1999, BIG comprised members from all seven villages surrounding the wetland. It had established byelaws affecting irrigation water use and cultivation as well as controls for pests and diseases and the use of farm inputs.

Perceptions of the environmental impacts of wetland utilization

The biophysical characteristics of wetlands underpin both the potential for, and constraints to, wetland resource use by communities. Ill-considered and poorly planned wetland development can undermine the human benefits derived from wetlands, often with a consequent downward spiral of increasing poverty and environmental degradation (Falkenmark *et al.*, 2007). The extent to which communities perceived environmental change as a consequence of wetland utilization was investigated. In Kilombero, a range of environmental issues were identified by people. The principal cause of concern, and one recognized by the vast majority of households, was more rapid drying of wetland soils than in the past. This was attributed to climatic changes, but villagers also believed that it was caused by land-use practices, associated with agricultural land preparation within the wetland (e.g. tilling). Only a small minority of households acknowledged that utilization of the wetland had an impact on downstream users and, where this was recognized, the primary issue related to impacts on water quality and sediment load, rather than on water quantity. It is possible that this does not reflect true understanding but rather local people feigning ignorance. This is not uncommon in situations such as this where people do not want to acknowledge the implications of their actions (Scott, 1985).

In the Bumbwisudi study, a number of environmental issues were identified, including soil erosion in the catchment, attributed to livestock keeping, poor farm management and bush clearing. Many villagers complained of declining soil fertility within the wetland and believed that this was best solved by increased fertilizer use. Analyses of aerial photographs showed that in recent years there have been considerable changes in the land cover of the wetland and its surroundings. These analyses indicated that between 1977 and 1989 significant declines in dense vegetation cover and mixed cropping, on the interfluves, were associated with comparable increases in sparse vegetation and paddy cultivation in the wetland. There have been no formal studies, but discussions between the study team and villagers and Department of Environment officials indicated that populations of frogs, butterflies, grasshoppers, lizards and birds (e.g. grey heron, black shouldered kite, white browed coucal, wood owl, green wood hoopoe and wood peckers) had all declined in recent years. There was no quantitative information on the number of villagers who are concerned by these changes, but generally environmental concerns were believed to be low.

At Chipala Ibenga, downstream households reported reduced stream flow after cessation of rains and increased periods of the stream drying after the rainy season. Most households complained of pollution of water from human faeces, domestic waste and sediments. Such pollution was said to be prevalent at the onset of the rains, causing a number of health problems including dysentery and diarrhoea. Ecological concerns such as deforestation and erosion in the area surrounding the wetland were not of serious concern. At Luano, future concerns related primarily to growing pressure arising from increased population leading to increased conflict over wetland access, rather than issues of environmental degradation *per se*.

Conclusions

The research conducted highlighted the fact that largely in the absence of assistance from the national government and with disparate polices and legislation, wetlands in both countries are utilized for agriculture and play a crucial role in the provision of basic needs and household survival. In all cases the wetlands made an appreciable contribution to food security, household income and welfare, both through provision of natural resources and through utilization for agriculture. However, it is clear that socio-economic differentiation leads to significant disparity in the agricultural utilization of the wetlands and the benefits derived from them. Both the absolute, and the proportion of, household income generated from wetlands are highly site-specific but also related to socio-economic status. In the majority of villages considered, a significant proportion of all households (irrespective of socio-economic status) utilize the wetlands in coping strategies during times of food scarcity. In all cases, control of wetland resources is determined by a complex amalgamation of modern and traditional institutions and arrangements. The four case studies provide largely qualitative indicators that increasing wetland utilization is having some negative environmental impacts. However, although communities identified a wide range of site-specific biophysical constraints (e.g. lack of water and poor soil fertility) and socio-economic constraints (e.g. lack of markets and high labour requirements) to the benefits to be gained from wetlands, environmental concerns were not high amongst them.

Developing appropriate policies for sustainable wetland utilization is not easy. The complex nature of livelihoods and their relationships to linked systems of natural resources make it difficult to identify and define authority structures that can take overall responsibility for wetland resource use and management. Furthermore, enforcement of formal regulations for wetland use and agriculture will not succeed in countries, like Tanzania and Zambia, that lack the resources to monitor and police wetland utilization. Pragmatically it seems that only those policies that really empower local people to manage and control the natural resources in their own landscape will be successful. To be effective such policies need to: 1) ensure equity by preventing community elites capturing resources; 2) bring immediate as well as long-term benefits to farmers; 3) promote cooperation and unified action at many levels; 4) ensure the maintenance of a wide array of ESS; and 5) bring about largely self-regulating and self-enforcing incentives for sustainable management. Currently, such policies remain elusive in Tanzania and Zambia. Although CBNRM strategies are increasingly perceived as an appropriate way forward, they must be considerably strengthened, if continuing wetland loss, to the detriment of local livelihoods, including smallholder farmers and society at large, is to be countered.

Acknowledgements

The case studies were collaborative initiatives undertaken between the Food and Agriculture Organization of the United Nations (FAO), the International Water Management Institute (IWMI) and IUCN-ROSA (the International Union for the Conservation of Nature – Regional Office for Southern Africa). They were funded by the FAO Netherlands Partnership Program. The author is grateful to Mutsa Masiyandima and Barbara van Koppen for assistance and to the case study partners who conducted a lot of the fieldwork: A. Daka, I. Ntambo, M. Kaemba and R. Moyo in Zambia; and S. Shaaban, M. Mchenga, A. Mbinga, R. Kangalawe, E. Liwenga, P. Mafuru, M. Futakamba, L. Simukanga, C. Tandari and K. Chisara in Tanzania. The involvement of the communities that participated in these studies and especially those households who gave up their time to complete the lengthy questionnaire is gratefully acknowledged.

References

Adams, M. (2003) *Land tenure policy and practice in Zambia: issues relating to the development of the agriculture sector*, Mokoro Ltd, Oxford.

Adekola, O., Morardet, S., de Groot, R. and Grelot, F. (2008) 'The economic and livelihood value of provisioning services of the Ga-Mampa wetland, South Africa', World Water Congress 2008, Montpellier.

Bakobi, B. L. M. (1993) 'Conservation of wetlands of Tanzania', in G. L. Kamukala and S. A. Crafter (eds) *Wetlands of Tanzania, proceedings of a seminar on wetlands of Tanzania*, Morogoro, Tanzania, 27–29 November, 1991, IUCN, Gland.

Bell, M., Faulkner, R., Hotchkiss, P., Lambert, R., Roberts, N. and Windram, A. (1987) *The use of dambos in rural development, with reference to Zimbabwe*, Loughborough University and University of Zimbabwe. Final report to ODA, p. 235.

Bikangaga, S., Picchi, M. P., Focardi, S. and Rossi, C. (2007) 'Perceived benefits of littoral wetlands in Uganda: a focus on the Nabugabo wetlands', *Wetlands Ecology and Management*, vol. 15, no. 6 , pp. 529–535.

Bryceson, D. (1998) *De-agrarianisation and income diversification in sub-Saharan Africa*, Presentation to the De-agrarianisation and rural employment workshop, Centre for Research and Documentation (CDR), Kano, Nigeria.

Campbell, B. (ed.) (1996) *The miombo in transition: woodlands and welfare in Africa*, Centre for International Forestry Research (CIFOR), Bogor, Indonesia.

Emerton, L. (2005) *Values and rewards: counting and capturing ecosystem water services for sustainable development*, IUCN Water, Nature and Economics Technical paper no. 1, IUCN Ecosystems and Livelihoods Group, Asia.

Falkenmark, M., Finlayson, C. M. and Gordon, L. (2007) 'Agriculture, water, and ecosystems: avoiding the costs of going too far', in D. Molden (ed.) *Water for food, water for life: a comprehensive assessment of water management in agriculture*, Earthscan, London.

FAO (1998) *Wetland characterization and classification for sustainable agricultural development*, Proceedings of a sub-regional consultation, Harare, Zimbabwe 3–6 December 2007.

FAO (2001) Socio-economic and gender analysis (SEAGA) Programme Field Handbook 2 (Vicki Wild ed., 2001), http://www.fao.org/sd/seaga/downloads/En/FieldEn.pdf (accessed 23rd October 2012).

Government of the Republic of Zambia (undated) *Environmental Council of Zambia: Zambia wetlands strategy and action plan (ZWSAP)*, Lusaka, Zambia.

Government of the Republic of Zambia (2008) *Ministry of Tourism, Environment and Natural Resources, Country Programme Strategy, 2008–2010*, Lusaka, Zambia.

Government of the Republic of Zambia (2009) *Ministry of Tourism, Environment and Natural Resources, National Policy on Environment*, http://zm.chm-cbd.net/implementation/legislation/policies-related-environment-and-biological/national-policy-environment-npe/npe_main_body_2009.doc (accessed 23rd October 2012).

Kamukala, G. L. and Crafter, S. A. (eds) (1993) *Wetlands of Tanzania*, Proceedings of a seminar on wetlands of Tanzania, Morogoro, Tanzania, 27–29 November, 1991, IUCN, Gland.

Kangalawe, R. Y. M. (2001) *Changing land-use patterns in the Irangi Hills, central Tanzania: a study of soil degradation and adaptive farming strategies*, PhD dissertation no. 22, Department of Physical Geography and Quaternary Geology, Stockholm University.

Kikula, I. S. (1997) *Policy implications on environment: the case of villagisation in Tanzania*, The Nordic Africa Institute, Uppsala, Sweden.

Kokwe, M. and Mulolani, D. (2003) *Zambia country paper on wetlands, biodiversity and livelihoods*, Unpublished report to IWMI, Colombo.

Liwenga, E. T. (2003) *Food insecurity and coping strategies in semiarid areas: the case of Mvumi in Central Tanzania*, PhD Thesis, Department of Human Geography, Stockholm University.

Lwoga, C. M. F. (1985) 'Seasonal labour migration in Tanzania: the case of Ludewa District', in G. Standing (ed.) *Labour circulation and the Labour process*, Croom Helm, London.

Matthews, G. V. T. (1993) *The Ramsar Convention on wetlands: its history and development*, Ramsar Convention Bureau, Gland.

McCartney, M. P., Morardet, S., Rebelo, L. M., Finlayson, C. M. and Masiyandima, M. (2011a) 'A study of wetland hydrology and ecosystem service provision: GaMampa wetland, South Africa', *Hydrological Sciences Journal*, vol. 56, no. 8, pp. 1452–1466.

McCartney, M. P., Rebelo, L. M., Mapedza, E., de Silva, S. and Finlayson, C. M. (2011b) 'The Lukanga Swamps: use, conflicts and management', *Journal of International Wildlife Law and Policy*, vol. 14, nos 3–4, pp. 293–310.

McCartney, M. P., Rebelo, L. M., Senaratna Sellamuttu, S. and de Silva, S. (2010) *Wetlands, agriculture and poverty reduction*, IWMI Research Report 137, IWMI, Colombo.

Millennium Ecosystem Assessment (MA) (2005) *Ecosystems and human well-being: wetlands and water synthesis*, World Resources Institute, Washington, DC.

Mukanda, N. (1998) 'Wetland classification for agricultural development in Eastern and Southern Africa: the Zambian case', in FAO (ed.) *Wetland characterization and classification for sustainable development*, Proceedings of a Sub-Regional Consultation, 3–6 December 1997, Harare, Zimbabwe.

Nkhata, B. A. and Breen, C. M. (2010) 'Performance of community based natural resource governance for the Kafue Flats (Zambia)', *Environmental Conservation*, vol. 37, no. 3, pp. 296–302.

Owen, R., Verbeek, K., Jackson, J. and Steenhuis, T. (eds) (1995) *Dambo farming in Zimbabwe: water management, cropping and soil potentials for smallholder farming in the wetlands*, University of Zimbabwe Publications, Harare.

Ramsar Convention Secretariat (2007) *National wetland policies: developing and implementing national wetland policies*, Ramsar handbooks for the wise use of wetlands, 3rd edition, vol. 2, Ramsar Convention Secretariat, Gland.

Rebelo, L.-M., McCartney, M. P. and Finlayson, C. M. (2009) 'Wetlands of sub-Saharan Africa: Distribution and contribution of agriculture to livelihoods', *Wetlands Ecology and Management*, vol. 18, no. 5, pp. 557–572.

Schuyt, K. and Brander, L. (2004) *The economic values of the world's wetlands*, WWF, Gland.

Scoones, I. (1991) 'Wetlands in drylands: key resources for agricultural and pastoral production in Africa', *Ambio*, vol. 20, no. 8, pp. 366–371.

Scoones, I., Chibudu, C., Chikura, S., Jeranyama, P., Machaka, D., Machanja, W., Mavedzenge, B., Mombeshora, B., Mudhara, M., Mudziwo, C., Murimbarimba, F. and Zirereza, B. (1996) *Hazards and opportunities – farming livelihoods in dryland Africa: lessons from Zimbabwe*, Zed Books Ltd, London and New Jersey.

Scott, J. C. (1985) *Weapons of the weak: everyday forms of peasant resistance*, Brevis Press, Bethany, Connecticut.

Woodhouse, P., Bernstein, H. and Hulme, D. (2000) *African enclosures? The social dynamics of wetlands in drylands*, James Currey, Oxford.

Zwarts, L., van Beukering, P., Kone, B. and Wymenga, E. (2005) *The Niger, a lifeline: effective water management in the Upper Niger Basin*, RIZA, Lelystad/Wetlands International, Sévaré/Institute for Environmental studies (IVM), Amsterdam/ A&W ecological consultants, Veenwouden, the Netherlands.

3 Catchments and wetlands

A functional landscape approach to sustainable use of seasonal wetlands in central Malawi

Adrian Wood and Patrick Thawe

Summary

Cultivation of seasonal wetlands makes an important contribution to rural livelihoods in many parts of Malawi. However, these wetlands are under increasing pressure due to population growth, the development of markets for wetland crops, and land degradation both in wetlands and in their catchments. These pressures are likely to increase with the challenge of climate change. A multi-community process of developing a consolidated understanding of wetland functioning, especially the links between wetlands and their catchments, and identifying specific innovations, which can help improve the possibilities of sustainable wetland use, is described in this chapter. It shows that over the generations rural communities have built up an understanding of wetland functioning to varying extents, and that when consolidated and supported by appropriate external scientific knowledge, this can provide a basis for community identification of specific measures that will improve the sustainability of wetland use. Together, the upland and wetland technical measures form a functional landscape approach (FLA) which takes a system-wide view recognising ecological and socio-economic processes. However, while technical measures may help address specific problems, the critical need is for the development of community institutions to ensure landscape-wide coordination of these measures in both wetlands and their catchments, as well as for economic incentives to increase the value of wetlands and make the application of the FLA worthwhile. Lessons from this experience feed into the debates about institutional development and economic motivation as critical elements of wetland management for sustaining livelihood benefits.

Wetland management and the functional landscape approach

Although wetland research and management has often focused on specific wetlands and the ecosystem services (ESS) (see Chapter 1) that they provide, there has been recognition and a growing understanding that specific sites are part of wider environmental systems and subject to system-wide processes, especially hydrological and geomorphological (Blumenfeld *et al.*,

2009; Ramsar Convention Secretariat, 2010). Indeed, it is these processes that cause wetlands to evolve, expanding in some cases and disappearing in others. This dynamic nature of wetlands, changing both spatially and temporally, has helped draw attention to these system-wide relationships and questions of how and from where water enters a wetland system, and how and where it leaves. In particular, the linkage between wetlands and their catchments, from where water and sediment are derived, and the influence of land use changes in catchments, such as deforestation and cultivation, are now well recognised as influencing the temporal and spatial availability of water in wetlands (Newson, 1997; Bossio *et al.*, 2010).

As with the evolution of water resource management into a multi-systems view, through river basin approaches and then integrated water resource management, it is clear that wetland management has to not only focus on the biophysical systems but also on the social, economic and political processes (Agol, 2010). These operate in both the catchments from where water and sediment are derived, as well as within the wetlands and the communities who are using these areas (Lenton and Muller, 2009). Hence sustainable wetland management has to address both biophysical and socioeconomic processes operating within the whole stream valley or river basin where a wetland is located.

The system-wide and multi-disciplinary scale of the perspective needed for sustainable wetland management draws attention to the functioning of the landscape, and the terrain in and around the wetland which together form the hydrological catchment. It is the maintenance of the functioning of this landscape in ways which are supportive of wetland ESS that are the basis for achieving sustainable wetland management. In this functioning it is important to recognise the mutual interactions and reciprocity between people and ESS across the landscape, through the way in which human management can support the functioning of ESS and the way the ESS support people (MA, 2005; Wood and van Halsema, 2008).

The term 'functional landscape' has been developed in ecology, particularly with respect to the habitat required to maintain particular fauna, often as part of a conservation and rehabilitation process (Forman and Godron, 1986). It has also been used to describe the alteration by communities of the lands around them to better serve their own needs by enhancing the survival of certain plants and reducing the presence of others that they find of no, or limited, use. The landscape scale of thinking has also become part of some wetland rehabilitation work with the term 'operational landscape unit' used with respect to connections of a hydrological and biotic nature between different parts of a wetland or former wetland which is being rehabilitated (Verhoeven *et al.*, 2008). Most recently, landscape approaches have been recognised as important for resolving conflicts between competing land uses and for improving the efficiency of overall resource use (EcoAgriculture Partners, 2012), rather in line with the views of the need to balance ESS presented by the Millennium Ecosystem Assessment (MA) and the Guidelines on Agriculture and Wetlands Interaction (GAWI) study (see Chapter 1).

Linking to these understandings, the term 'functional landscape approach' (FLA) was developed in 2005 through discussions by Wetland Action (WA) and local communities at the start of a rural livelihoods project. This idea was also applied in the subsequent Striking a Balance (SAB) project.[1] This term was seen as appropriate given the way the field projects sought to help communities improve their management of land use in both catchments and wetlands in ways that recognised their functional connections or inter-linkages. The FLA was seen to have the potential to sustain and enhance ESS and especially the livelihood benefits they provide across different facets of the landscape. Hence the FLA term recognises that wetland sustainability is based on biophysical processes not only in the wetlands but also in their catchments, and that these are influenced by human decision making and so are subject to institutional arrangements as well as economic incentives, social influences and cultural factors (see also Chapter 5).

Striking a Balance (SAB) project and the landscape in central Malawi

The concept of the FLA for sustainable wetland management has been applied through the collaboration of WA and MALEZA – the Malawi Enterprise Zone Association – in projects implemented in the Simlemba Traditional Authority (TA) in central Malawi. This work, which has been on-going since 2005, seeks to explore how to achieve sustainable wetland use for livelihood development, especially through sustainable wetland agriculture. It recognises that this requires maintaining a balance of ecosystem service uses within the wetlands and also improved land management in the catchments.

Central Malawi was chosen for the SAB project because it was known that wetlands play an important role in supporting rural livelihoods in this area. Indeed, in much of south central and southern Africa, seasonal wetlands have traditionally been used for dry season cultivation with small dimba gardens created, and residual moisture or small-scale hand/can irrigation used to produce food relish (mostly vegetables) and to supplement starch staple (maize) supplies (Trapnell and Clothier, 1937). In particular, seasonal wetlands have been used more intensively in drought years as a coping strategy and have been an important means by which people have survived (Sampa, 2008).

The importance of cultivation in seasonal wetlands has increased in this area, as in many parts of Africa, during the last few decades (see Chapter 1). This has been due to a shortage of upland rain-fed fields caused by population growth, the decline in the fertility of those fields and the impact of rainfall variability, especially periodic droughts, on the upland crops. There have also been growing commercial opportunities in the expanding trading and urban centres for the products of the wetland gardens, especially vegetables, and this has further stimulated wetland cultivation (Woodhouse *et al.*, 2000). Wetlands are expected to come under increasing pressure as the vagaries of climate change intensify, and the possibility of continuing secure harvests in wetlands may provide an adaptive strategy in the face of this challenge (Mortimore and Adams, 2001; Wood and Dixon, 2009).

While wetland cultivation in the dry season is expanding, there is evidence of wetland loss and degradation through gulley formation which lowers the water table and water storage, sediment deposition from the uplands which leads to soil deterioration and through the apparent general drying out of wetlands. Furthermore there is a long-standing debate about the sustainability of wetland cultivation, its impact on stream flow and upon wetland degradation (Faulkner and Lambert, 1991; von der Hyden and New, 2003).

While much of the debate about sustainable wetland cultivation has focused on dambos, seasonal headwater wetlands without permanent stream courses (Roberts and Lambert, 1990; McCartney, 2000), streams and river valleys and their wetlands are also subject to similar discussions about processes of gulley formation, the extent of water storage in these areas and the impacts of catchment deforestation and intensified cultivation (Shaxson, 2006; Blumenfeld *et al.*, 2009; Bossio *et al.*, 2010). There has also been wider discussion about wetland use and the need to balance the interests of cultivation, water supply and wildlife (Frenken and Mharapara, 2001).

While much of the recent research has confirmed that some cultivation in wetlands is possible without degrading them (Roberts, 1988), the hydrology of these areas, especially dambos, are still not properly understood, although the importance of catchments for water storage (within the landscape and as groundwater) is confirmed (Bullock, 1992).

Conclusions on three particular debates are of particular note here:

- Gulley formation is probably due to artesian water pressure, rather than surface flow, in some dambos with shallow clay lenses, but this is unlikely in stream valleys (McFarlane and Whitlow, 1990).
- Deforestation and hoe cultivation and its associated soil compaction in catchments leads to increased runoff rather than increased infiltration due to reduced evapotranspiration by trees (Shaxson, 2006).
- Traditional cultivation practices in the wetlands may have a limited impact on erosion and gulley formation, in contrast to major drainage ditches associated with commercial use of these areas (Roberts, 1988).

In the light of these issues, the SAB project sought to explore practical measures to reduce the risks of wetland degradation, enhance water storage and improve livelihoods. Further, the study engaged, to some extent, with the work relating to motivations for sustainable agricultural intensification as proposed by Boserup (1965) and supported from studies in the Machakos area of Kenya (Tiffen *et al.*, 1994).

Simlemba Traditional Authority (TA)

The Simlemba TA is located in the north-east of Kasungu District in the central region of Malawi (Figure 3.1). This is part of the headwaters of the Dwangwa River, a major tributary to Lake Malawi. The terrain is a gently rolling plateau with a few inselbergs, although the plateau is increasingly dissected in the east as the escarpment down to Lake Malawi is approached.

The area has an average annual rainfall of between 800 mm and 1,000 mm with the rain concentrated in the period from late November to March. The reliability of this is not known, but as with other parts of the country Simlemba suffered from four droughts in the first six years of the twenty-first century. In fact this area was known for its problems with water availability during colonial times (pre-1964), and that was reportedly one reason why settlement has been limited until recent decades (pers. comm., Francis Shaxson).

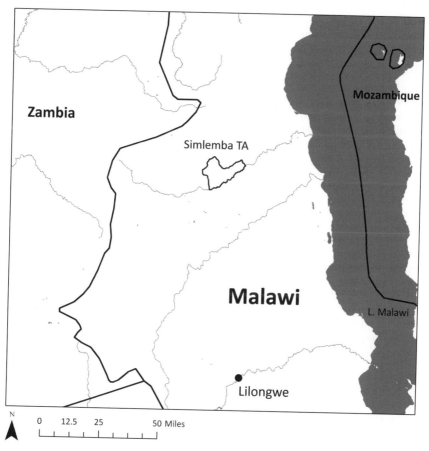

Figure 3.1 Map showing the location of the Simlemba Traditional Authority

The soils in Simlemba are mostly sandy loams of the latosol type (KDC, 2004). As in other parts of the country there are signs that these soils are deteriorating with a breakdown in their structure and declining fertility, while hard pans due to repeated hoeing are also reported, which in turn reduce water infiltration and increase runoff (Douglas *et al.*, 1999; Shaxson, 2006). There is some natural forest remaining in the hilly areas in the north and east of the TA and these areas are identified as unsuitable for cultivation (KDC, 2004).

According to the 2008 census, the Simlemba TA covered 251 km² and had a population of 29,400 giving it a population density of 117 persons per km² (GOM, 2008). This is 26 per cent greater than it was in 1998. This rapid population growth is in part the result of migration, the TA having experienced considerable in-migration into its eastern and southern areas by people from areas to the south and south-west, especially Mchinji District where land shortages are severe.

As in many parts of Malawi, rural poverty is severe with 49 per cent of the population of Kasungu District living below the poverty line in 1998 (KDPD, 2003). In October 2005, after a series of drought years, 30 per cent of families in the district were reported to be without any food reserves and over 700 (out of some 2,000) boreholes were dry, seriously increasing the burden of water collection on women (pers. comm., Kasunga District Planning Department (KDPD)). Other aspects of the development challenge are small farm size – 50 per cent of households had less than one hectare of farmland in 1998, and a high number of female-headed households – reported to be 34 per cent of all households in 1998 (KDPD, 2003). In this situation, the repeated focus of the District Development Plans in the last decade has been to address food insecurity through small-scale irrigation in the seasonal wetlands to produce crops for domestic use and sale (KDC, 2003, 2006, 2009). To this end a number of micro dams have been built and treadle pumps have been provided, with one or two obtained by most villages with wetlands.

Seasonal wetlands and their provisioning services

Seasonal wetlands are widespread in the Simlemba TA, while a very small number of permanent ones are also found. The majority of these seasonal wetlands are in stream valleys where the streams overflow during the rainy season across small floodplains. In most cases the streams themselves are seasonal and have no flow for six months or more. Only one or two streams are perennial in this area, these usually being found where the catchment is forested and protected from clearance and cultivation. In some cases, near the head of the drainage networks, there are dambos.

Wetland use is common but by no means universal. Of the 112 villages in the Simlemba TA, approximately 30 have wetlands within their territory that are suitable for cultivation. There is a tradition, based on family ties, of sharing wetlands for some uses, notably grazing and the collection of reeds or thatching grass, but cultivation of wetlands outside the village of residence was unusual until the last decade when upland crop failures led to it becoming more common.

These seasonal wetlands have been used traditionally for a range of ecosystem provisioning services (Table 3.1). The gender division of labour affects the involvement of men and women in the use of these services and also their prioritisation. The focus of men is on income generation activities while women prioritise domestic uses, although women are widely involved in wetland cultivation. This experience in the three SAB villages is shown in Table 3.1.

Use of these services has changed over the last four decades, with cultivation expanding and other uses declining. Traditionally only a few households used wetlands for cultivation on a regular basis, plots were small and often fenced to keep out livestock. Droughts from the 1980s onward, and especially more recently, and the recognition of market opportunities for wetland crops, have led to more households developing dimba gardens and for the size of these to

Table 3.1 Wetland uses, users and gender prioritisation: three villages

Wetland provisioning services	Users – age and gender	Priority order by men	Priority order by women
Water (from springs, hand-dug wells and streams) for drinking and domestic uses	Mostly women	2	1
Cultivation of crops for food and income (maize, vegetables, tomatoes, bananas, sugar cane)	All people, in all socio-economic categories	1	2
Collecting reeds for making mats	Mostly men and boys	5	–
Collection of thatching grass and grass for tobacco nurseries	Mostly men, some women	4	4
Grazing and water for livestock	Mostly men	3	–
Collecting sand for domestic use – construction, maize processing and cleaning pots	Women and girls	–	6
Collecting clay for pottery and smearing of houses	Women and girls	–	3
Fishing	Mostly boys, few men	6	–
Collecting edible wild plants (*chinaka* = edible tuber), fruits and relish	Women and girls	–	4

Source: Msukwa, 2007

increase. As a result the area of cultivation has risen from between 5 per cent and 20 per cent of a wetland to between 30 per cent and 90 per cent, depending on water availability. A converse trend has been the decline in wetland use for livestock grazing in most areas. This has been in part due to a decline in cattle numbers, caused by disease in the 1980s, but also by competition with farmers wanting to develop gardens. Those few wetlands with permanent water supplies have become a focus for the livestock from several villages.

Expanding wetland cultivation has had a negative impact on the collection of wild plants. The collection of clay from wetlands has also declined, in part because of the expansion of cultivation but also because of the increased use of plastic containers rather than clay pots. Despite the introduction of water pumps in some villages, many settlements in the Simlemba TA remain dependent on natural water sources. With the growing population and their demand for domestic water, and the increased use of water for wetland cultivation, especially once treadle pumps are used, the pressures on the water supplies available in the seasonal wetlands have increased.

The major wetland crops are green maize, beans and tomatoes, with vegetables (mostly leaf crops) and sugar cane also present in all villages (Table 3.2).

Of particular note is the fact that wetlands are used for cultivation by both male- and female-headed households and by households of all socio-economic groups (Table 3.3). Within these three villages, 88 per cent of all households were cultivating dimba gardens to some extent and 77 per cent of the households were regarded as high users of these areas (i.e. households using the wetlands, mainly for cultivation, from soon after the rainy season up to the onset of the next rainy season.) Amongst the male-headed households, only 12.4 per cent did not use dimba gardens, while only 9.6 per cent of female-headed households were in this category. In terms of 'high users', 70.1 per cent of female-headed households were in this category and 79 per cent of male-headed households. These figures suggest that there is no bias against wetland use for female-headed households. Despite the small sample there is a possible relationship between wealth categories (developed through community discussions and based on assets, income and food consumption) and wetland use, with only 4.5 per cent of the better-off households not using wetlands but 14.6 per cent of the poorest not using them. This may be due to limited labour or other resources available to the poor, or the use of kukapumba, migrating for work, as the way to address food shortages and poverty.

Wetland agriculture is now a major contributor to household income in the Simlemba TA for those villages that have wetlands and have been able to develop parts of them for cultivation. Overall in the three villages studied in detail, wetlands provided 37 per cent of the domestically used crops, while they accounted for 55 per cent of the farm-generated cash income.

Table 3.2 Crops grown in wetlands – ranking by area and value: three villages

Wetland crop	Priority use (for sale or domestic consumption)	Rank by area	Rank by cash income (for crops sold)
Maize	Domestic use and sale	1, 1, 2	1, 3, 5
Beans	Sale and domestic use	2, 2, 7	2, 4, 6
Tomatoes	Mostly for sale, little for domestic use	3, 4, 3	3, 1, 1
Onions	Mostly for sale, little for domestic use	7, 6	7, 2
Vegetables	Domestic use and sale	5, 3, 5	4, 7, 2
Irish potatoes	Domestic use, little for sale	8, 5	5, 6
Sugar cane	Sale and domestic use	4, 7, 1	6, 5, 3
Sweet potato	Domestic use and sale	6	8
Rice	Domestic use	9	9
Bananas	Domestic use and sale	4	4

Source: Msukwa, 2007

Table 3.3 Wetland-using households by socio-economic status: three villages

User group	Poorest MHH	Poorest FHH	Medium MHH	Medium FHH	Better off MHH	Better off FHH	No.	%
Non-user	3	3	8		1		15	11.7
Low users			2		1		03	2.3
Medium users		4	5	2			11	8.6
High users	20	11	37	11	20		99	77.3
Total	23	18	52	13	22	0	128	
Total %	18.0	14.1	40.6	10.2	17.2	0	100	
M %	23.7		53.6		22.6		97	
F %		58.1		41.9			31	

MHH – Male Headed Households; FHH – Female Headed Households
Source: Msukwa, 2007

Beyond these summary figures it should be recognised that wetland crops can contribute in multiple ways to the well-being of many households. Domestic food security is improved, while diversity of diet may improve, especially during the late dry season and the early rains when other upland crop harvests have been used. Income from wetland crops is used for the purchase of other food and households goods as well as for the payment of school fees, while seeds and inputs for upland crops are often purchased with these funds. Some households have even invested funds from wetland agriculture in businesses, such as small shops, or farm enterprises, such as raising chickens or pigs (Ndiyoi *et al.*, 2009).

Given this growing importance of wetlands, there is a widespread desire amongst the wetland-using communities to explore ways in which they can further develop their seasonal wetlands to secure their well-being and improve their livelihoods, as well as preventing the degradation and loss of the ESS in these areas.

Local knowledge of wetland dynamics

Exploring the existing knowledge of wetlands in the communities in Simlemba was undertaken in 2005 at the start of the process to identify measures to improve wetland management (Wood, 2005). Local knowledge was formalised through extensive discussion with the communities in four villages who reviewed the state and functioning of their wetlands during transect walks and subsequent focus group discussions and a multi-community workshop. These showed that there was a considerable body of knowledge about the wetlands and their dynamics, presumably as a result of the various uses of wetlands by the farmers, both male and female, over generations. However, this knowledge varied from one community to another, was based on specific experiences and no village had built up a comprehensive understanding. Nonetheless, when the knowledge across the various villages was compiled, a fairly complete picture of wetland dynamics was created which, with some limited scientific input, could inform community-based planning to enhance the sustainable use of these areas.

The results from the transect walks and community discussions showed that the key understandings within this local knowledge were:

- linkages between catchments and wetlands were recognised, although these were seen primarily in terms of the deposition of eroded material in the wetlands which caused a deterioration in the quality of the soils at the edge of these areas;
- the moisture in the wetlands during the dry season was seen to be a result of rainfall and flood water being stored in the valley sediments, but in some cases it was recognised as being linked to water stored in the catchments which was seen to contribute to the groundwater in the wetland;
- considerable concern was expressed about the development of gulleys in the seasonal wetlands and their negative impact on water availability in the areas adjacent to them where the water table was lowered;
- rapid runoff from the catchments was seen as a possible contributor to erosion in the wetlands and as a source of the powerful water flow and turbulence that was associated with the formation of gullies;

- competition for water in the dry season by farming in parts of the wetlands with intense use was recognised and actions were already being taken to reduce extraction using treadle pumps after five months of the dry season in order to conserve water for domestic use in the driest time of the year; and
- lowering of the water table was reported from wetlands where the area of sugar cane cultivation had been expanded, and related negative impacts were noted on water availability for vegetable growing.

Other issues that were raised in the discussions during the workshop are based on local experience and some external scientific knowledge contributed by the facilitator. These included:

- potential impacts on water availability from the cultivation of eucalyptus trees grown at the margins of the wetlands and within them;
- possible turbulence and consequent erosion caused by the excavation and clearance of vegetation around the large crater-like, hand-dug wells, which provide 'walk-in' access to irrigation water sometimes several metres below the surface, located in the centre of wetlands where water movement is most rapid during the floods;
- value of soil and water conservation measures, composting, agroforestry and afforestation (with native tree species, not eucalyptus) on reducing runoff from the uplands/catchments into the wetlands and the impact of some of these measures on improved water infiltration which will benefit the wetlands;
- possibilities for improved efficiency in the use of irrigation water, through greater care with treadle pump spraying of water, as well as improved timing of irrigation after the hottest part of the day so as to make this resource last longer into the dry season;
- need for coordinated water management to drain parts of permanent wetlands in some seasons, but also to avoid over-drainage and erosion; and
- the importance of improving market opportunities for wetland crops in order to gain maximum value from these areas.

Overall there was a feeling that with the growing pressures upon the wetlands, less water was going into these areas but more was being taken out of them. Certainly there had been lower water tables in the wetlands during the dry season, although the influence of the recent droughts should not be ignored.

Towards a functional landscape approach for sustaining wetland use

The multi-community and multi-stakeholder meeting, which included some government extension staff as well as farmers, consolidated this understanding from the different communities about wetland dynamics.

It tentatively developed the functional landscape conceptual framework to guide field activities, while a series of specific measures were identified for testing by farmers in order to explore how the goal of enhanced, but sustainable, wetland development and use could be achieved.

The FLA came about in these discussions from the evolving understanding amongst farmers of the links between wetlands and their catchments, and between upstream and downstream parts of stream valleys. The main form of linkage in the functional landscape was seen to be through water flows and the various hydrological processes, especially water infiltration and water erosion. The other major component of the FLA was an understanding of the way vegetation cover and land management can control water flows and associated erosion, and so affect water infiltration and water storage in soil and sub-strata in the catchments. Within this framework, two groups of technical measures were identified by the participants, some relevant to the seasonal wetlands and others relevant in the catchments (Thawe, 2008).

Wetland measures

The initial thinking about how to improve wetland management was based on zoning of uses within the wetland. A key factor driving this was the recognition that the centre of the wetland was where the water moved most rapidly during the floods and could do most damage. As a result it was suggested that this area should be protected from cultivation and wells should not be located here. The former was to reduce the presence of unprotected soil in gardens in the centre of the wetland, while wells and the human impacts around them were thought to facilitate gulley formation. The idea of not cultivating in the centre of wetlands was also probably stimulated by the government extension advice which is to protect stream course edges from erosion by preventing cultivation within 10 m of them.

Another aspect of the zoning approach was the proposal to maintain natural vegetation across the full width of the wetland at some locations in order to slow floods, stabilise the wetland soils and so reduce risks of erosion. These areas of natural vegetation may be economically valuable if they include medicinal plants or provide relish, craft products and opportunities for grazing, while they can also become a biodiversity reserve by providing natural habitat. Conversely, areas of natural vegetation can become refuges for pests, although they may also provide habitats for pest predators and assist in integrated pest management.

Maintaining or developing areas of natural swamp at the centre of a wetland were also seen as a possible way to help enhance ESS and the provision of natural products, while developing fish ponds was also seen as a way to maintain the mix of ESS and add value to the wetland. These ideas are in line with the ideas coming from the MA and GAWI work (MA, 2005; Wood and van Halsema, 2008; see also Chapter 1).

To complement the above measures which seek to enhance water storage, water extraction was another area identified where it was proposed measures should be taken. In particular, it was suggested that the extent of sugar cane cultivation in seasonal wetlands should be limited to reduce the extraction of water by these plants. Similarly it was proposed that eucalyptus trees should be removed from around the edge, and within, seasonal wetlands because of their reported impacts on the water table. A further aspect of controlling water extraction, already identified by some communities, was the development of community rules or agreements to limit or even halt the use of treadle pumps so as to control water extraction in the late dry season.

At the micro scale it was suggested that cultivation methods could be improved in order to reduce demands for water and the risks of erosion, as well as also improving the quality and value of the crops produced. Specific measures proposed included:

- limiting plot sizes so that areas with disturbed soils do not dominate the wetland (i.e. exceed 50 per cent) while areas of natural vegetation are maintained to help prevent erosion;
- adjusting the height of the cultivation beds relative to the wetland surface to minimise the need for drainage (by using raised beds in the post-rainy season time) or irrigation (by using basins or depression beds);
- using organic manure to help maintain soil fertility and remove the risks of increasing soil acidity and the need for liming;
- appropriating the spacing of plants to improve the quality of produce and to reduce the risks of disease;
- using crop rotation and intercropping to reduce disease risks and maintain soil fertility; and
- watering in the late afternoon, rather than mid-morning, to reduce the loss of water by evaporation, concentrating water in the beds rather than spreading it onto paths, as well as beds, and using depression beds or basins to hold the water.

Catchment measures

The critical issue with respect to the catchment in order to improve wetland sustainability was seen to be the need to enhance water infiltration into these areas so as to improve the flow of water through the subsoil strata and into the wetlands to provide long-term water recharge, especially in the dry season. Specific measures to assist this were identified as being:

- application of conservation farming and soil and water conservation measures in the upland fields in order to improve infiltration and to reduce runoff and erosion, which at the same time will help increase upland field yields;

- use of agroforestry techniques within fields and afforestation on non-farm land, with local species and some exotics, but excluding euca-lyptus, to reduce runoff and erosion; and
- maintenance of a 5–10 m-wide buffer zone of natural vegetation on the lower slopes of the catchments to control sediment and surface water movement into wetlands, thereby providing a last opportunity for water infiltration within the catchment.

The overall result of combining such measures across a stream valley might appear as shown in Figure 3.2. The upland practices should extend throughout the catchment up to the watershed of each basin, while the wetland practices should be applied in those areas of the wetland where cultivation is present.

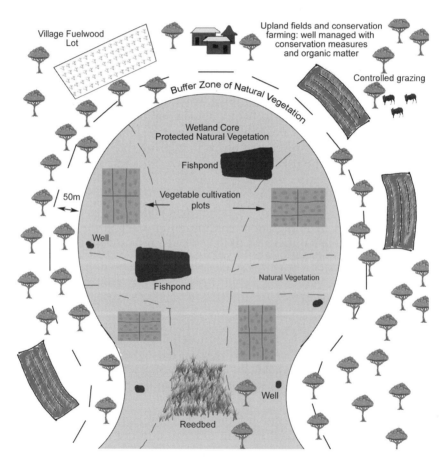

Figure 3.2 A model functional landscape, linking catchments and wetlands

Managing wetlands for ecosystem service benefits: institutional developments and economic incentives

While technical measures have the potential to improve the productivity and sustainability of seasonal wetlands, their implementation must be based on a common understanding of their value and so have widespread support in the community. In particular, there must be a means to ensure, or encourage, the adoption of such guidance. This discussion raised issues within the three pilot communities in the SAB project about institutions and incentives.

With respect to the institutional basis for wetland and catchment land management, it soon became clear that some form of coordination of land use practices and authority to implement changes was required as it could be necessary to control land uses by individual households and even relocate land rights. The existing village development committee (VDC) was seen initially as a body that could undertake this task and then advise the village headman who would use traditional structures to enforce any measures seen as necessary. However, after some trials and discussion the communities, through the local non-governmental organisation (NGO), MALEZA, chose to develop a specific sub-committee of the VDC in order to address such land management issues. This became the Village Natural Resource Management Committee (VNRMC) and it developed byelaws to provide guidance across the communities. These regulations were approved by the village headman and the group village headman above him on behalf of the Chief who heads the TA (Sampa *et al.*, 2008).

The VNRMCs were not an initiative of the SAB project, but rather an output of the Forest Act (1997) and the Community Based Forest Management Act (2003) which had endorsed the use of community-based institutions for natural resource management (NRM) and the development of byelaws to ensure sustainable utilisation of forest resources. Empowering these institutions and orienting them to address wetland and catchment management issues thus became necessary. This has been achieved to a considerable degree and they have spread local wetland knowledge and built adaptive capacity to improve the management of wetlands and catchments within, and even outside, their communities. Byelaws that these VNRMCs have developed differ slightly between sites, but in general they include:

- designating a 5–10 m-wide buffer zone from the centre of the wetland or stream channel, in which no cultivation is allowed;
- ensuring livestock are always supervised in the wetland;
- prohibiting the removal of indigenous trees from within the wetland;
- advising on the planting of crops in beds which are like basins (depressions) in the hot season to use water efficiently;

- prohibiting the planting of eucalyptus trees in or near the wetland;
- limiting the area of sugar cane in sites where water is scarce;
- maintaining a non-cultivation zone of 5–10 m width with natural vegetation around the wetland;
- protecting afforested areas and zones of natural vegetation from fires, including the use of firebreaks; and
- limiting use of treadle pumps when domestic water supplies are affected by lowered water tables in the late dry season.

Widening the byelaws to take a holistic FLA is an on-going process for the VNRMCs as the catchment land management is seen as a critical contribution towards wetland sustainability.

With respect to incentives to motivate communities to value their wetlands more and so seek to improve management of them and their catchments, the main initiative has been to help farmers produce higher value crops. This work has focused on improving the quality of the crop through better cultivation techniques and establishing better market linkages through value chain development. This is an on-going process with crop changes occurring in response to market opportunities. For instance, beans were replacing vegetables and tomatoes in 2011 because of produce losses with the more perishable crops, and the ability to store beans for sale when the price is high.

Discussion of experience with the functional landscape approach

Experience has varied across the three pilot communities who have applied the FLA with support from MALEZA. This has depended on community coherence and the ability of the VNRMC to be effective, the availability of water for increased wetland crop production, and finally access to markets and hence the economic incentives for improved wetland and catchment management. The VNRMC has been weak in one community due to local political issues, water has been scarce in another village, while in the third two good market links and a strong VNRMC have led to more activity by farmers and the community as a whole.

Variations have also occurred in terms of the application of different elements of the FLA. In all cases communities were most interested in improving management of the cultivation within the wetlands, as that was seen as likely to give an immediate return through better-quality crops. Similarly, efficiency of water use and control of treadle pumps in the late dry season was supported because it was seen as likely to extend the cultivation season – even if it would be on a smaller scale – and hence extend food production and increase income for the majority. Prevention of late dry season use treadle pumps was achieved in one community despite the village headman being the major user of this technology.

In general, protecting the core of the wetlands from cultivation was agreed and implemented, the only exception being where the wetland was permanent and cultivation involved water level management through a network of draining channels leading to a central drain. Leaving contiguous plots of natural vegetation land across the whole width of the wetland was less successful, especially where land for dimba gardens was limited and more than 75 per cent of the dambo was cultivated.

The creation of a non-use infiltration zone around the wetlands was agreed and applied in all communities as they were well aware of how this could reduce the sediment deposition in the wetlands. This was achieved despite the fact that a few farmers had to relocate some of their fields at one site.

In general there was less enthusiasm for addressing the sugar cane and eucalyptus issues which involved individuals being challenged over the external impacts of their individual enterprises upon the community's wetland resource.

Despite the field discussions and inter-community meeting which showed a clear appreciation of the catchment–wetland linkages, the level of adoption of improved land management in the catchments was limited. Adoption of the soil and water conservation measures, agroforestry and afforestation, for which technical support was available from MALEZA, was patchy. Apparently the farmers who took up these measures did so because they saw immediate benefits for their upland crop, not because of any expected impacts in the wider landscape, notably the wetlands, which they may, or may not, have used.

This experience with applying the FLA raised two important issues that have to be addressed. First the benefits of the FLA will be maximised when this is applied across a complete stream/river valley, including all the catchment so as to improve infiltration and reduce runoff. However, this requires the coordination of land use and practices of land management across several villages, each with their own interests and different levels of understanding of the wetland dynamics, as well as differences in their institutional capacity. Developing and applying the potential authority of the group village headmen (responsible for several villages) and the Chief responsible for the TA will need to be explored in this respect.

The other issue is that as the FLA ideas are applied more widely, especially the attractive ones that support increased wetland crop production, demands upon the wetlands will increase. This will lead to pressures on the communities institutions, the VNRMC especially, and on the understanding in communities of the limits to wetland use to ensure sustainability. Hence, the need for strong local management, through the VNRMCs and the village headmen, will grow. Their ability to manage such situations will need to be built on a thorough understanding of the wetland dynamics, by combining local and scientific knowledge, and on management practices that can support an increased number of wetland users in a sustainable way.

The understanding of the importance of wetland and catchment management and the need for an FLA should be supported from several sections in Simlemba society, not just by those engaged in farming. Women should support this because of the improved recharge in the dry season of the wetland sediments in which shallow wells are dug for domestic use. The TA authorities should support the FLA as this will not only improve food security and help sustain improvements in income, but it may help by reducing catchment runoff and flood peaks and so minimise damage to bridges and other infrastructure caused by floods.

Conclusions

Application of the FLA in the Simlemba TA suggests that it could be possible to achieve sustainable use of these seasonal wetlands with increased livelihood benefits. Through measures in both the wetlands and the catchments, the ecological characteristics can be maintained and enhanced in some parts of the wetland to help retain water and provide the critical ESS (such as support and regulatory ones – see Chapter 1) that are needed to maintain wetland functioning. In other parts of the wetlands, based on the functioning of these ESS, provisioning services can be developed to address food security and economic needs. Overall within a wetland, or a system of wetlands along a stream valley, a balance of ESS can be maintained which will help ensure the sustainability of the wetlands and the people dependent on them, as the MA and GAWI studies suggest (MA, 2005; Wood and van Halsema, 2008).

While the functioning of the wetlands can probably be maintained through the mix of land uses and the management practices proposed for the wetlands and the catchments (see Chapter 1), economic and institutional sustainability must also be actively addressed. This will be through the improved quality of crops and market linkages on the one hand, and the building up of the VNRMCs and community support for them on the other. Essential for the institutional sustainability is the ability of the VNRMCs to build on the local knowledge of wetlands in the communities and interrogate and enhance this with external scientific knowledge so that there is a thorough and accurate understanding of wetland dynamics to support and justify byelaw development.

Overall, the balance of land uses that are sought through the FLA provide a way in which the traditionally competing and conflicting demands upon wetlands can be resolved in a mutually beneficial way (Wood and van Halsema, 2008; EcoAgriculture Partners, 2012). This can apply not just at the community level but also at the district and government level where the FLA provides the basis for a dialogue between the different interest groups, such as those concerned with conservation, water supply and food security (Wood, 2011). This discussion also provides the basis for trade-off discussions about different wetland uses as pressures increase (McCartney *et al.*, 2005; see also Chapter 11).

The FLA links to other discussions about NRM and sustainable agricultural intensification, especially that by Boserup (1965) and the more recent testing of these ideas in Kenya (Tiffen *et al.*, 1994). In the case of both wetland and drylands it is clear that more people may not always lead to more erosion. However, the key issues appear to be the understanding of resource dynamics by the community, the development of appropriate management institutions and the existence of appropriate market incentives.

This work from Simlemba provides one example of the way local knowledge can be supported with scientific knowledge to develop a combined understanding of natural resources and to develop from this a series of interventions to test (Raymond *et al.*, 2010; Sandbrook, 2012). The experience to date suggests that the FLA may be more widely applicable, but this has to be tested in different situations, with diverse environmental conditions – especially with respect to hydrology – and also varying socio-economic, institutional and cultural situations. While there is an inherent logic in the concepts of the FLA, and these are supported by the experience to date in Simlemba, adapting this process of interrogating local understanding, building an understanding of how specific wetlands work and developing interventions to be applied across the landscape will certainly be necessary in each situation. Documenting and learning from each case is the way to explore how the FLA can help address the almost certainly growing pressures on seasonal wetlands in Malawi and much of sub-Saharan Africa (SSA) as a result of the on-going environmental and economic changes, and especially the expected challenges of climate change.

Note

1 The SAB project was developed and implemented by WA in association with Harvest Help (a UK NGO now part of Self Help Africa) and MALEZA. It was funded by Netherlands Ministry of Foreign Affairs (DGIS) through Wetlands International (WI) and was part of that organisation's Wetlands and Poverty Reduction Project (WPRP). Specific data in this paper is drawn from three villages where pilot field activities have been undertaken. Other material in this chapter is based on six transect walks, eight group discussions with other wetland-using communities and a multi-community workshop undertaken in 2005 by WA for the Simlemba Rural Livelihoods Project of Harvest Help and MALEZA, funded by the Big Lottery (UK).

Acknowledgements

The authors wish to record their appreciation to the rural communities in Simlemba who contributed their knowledge and experience, as well as their time, in the discussions that led to this paper. They still face many challenges but hopefully some of the lessons and contributions from the SAB and Simlemba Rural Livelihoods project have helped them improve their livelihoods. The engagement with Harvest Help and MALEZA was critical and vital for this work and this is acknowledged with grateful thanks. The SAB project funding from the Dutch Ministry of Foreign Affairs through the WPRP of WI is also gratefully acknowledged by WA.

References

Agol, D. A. (2010) *Exploring Knowledge Interfaces for Integrated Water Resources Management: Findings from the River Nyando, Lake Victoria Basin, Kenya*, Ph.D. Thesis, University of East Anglia.

Blumenfeld, S., Lu, C., Christophersen, T. and Coates, D. (2009) *Water, Wetlands and Forests: A Review of Ecological, Economic and Policy Linkages*, Secretariat of the Convention on Biological Diversity and Secretariat of the Ramsar Convention on Wetlands, Montreal and Gland, CBD Technical Series no. 47.

Boserup, E. (1965) *The Conditions of Agricultural Growth*, Aldine, Chicago.

Bossio, D., Geheb, K. and Critchley, W. (2010) 'Managing water by managing land: addressing land degradation to improve water productivity and rural livelihoods', *Agricultural Water Management*, vol. 97, no. 4, pp. 536–542.

Bullock, A. (1992) 'Dambo hydrology in southern Africa: review and reassessment', *Journal of Hydrology*, vol. 134, nos 1–4, pp. 373–396.

Douglas, M. G., Mughogho, S. K., Saka, A. R., Shaxson, T. F. and Evers, G. (1999) *Malawi: An Investigation into the Presence of a Cultivation Hoe Pan under Smallholder Farming Conditions*, FAO, Rome (Occasional Paper Series no. 10, Investment Centre).

EcoAgriculture Partners (2012) *Landscapes for People, Food and Nature: The Vision, the Evidence and Next Steps*, EcoAgriculture Partners on behalf of Landscapes for People, Food and Nature Initiative, Washington, DC.

Faulkner, R. D. and Lambert, R. A. (1991) 'The effects of irrigation on dambo hydrology: a case study', *Journal of Hydrology*, vol. 123, nos 1–2, pp. 147–161.

Forman, R. T. T. and Godron, M. (1986) *Landscape Ecology*, John Wiley, New York.

Frenken, K. and Mharapara, I. (2001) *Wetland Development and Management in SADC Countries*, FAO, Harare, Zimbabwe.

GOM (Government of Malawi) (2008) *2008 Population and Housing Census, Preliminary Report*, Central Statistical Office, Zomba.

KDC (Kasungu District Council) (2003) *Kasungu District Development Plan*, KDC, Kasungu.

KDC (2004) *Kasungu District Environment Report*, KDC, Kasungu.

KDC (2006) *Kasungu District Development Plan*, KDC, Kasungu.

KDC (2009) *Kasungu District Development Plan*, KDC, Kasungu.

KDPD (Kasungu District Planning Department) (2003) *Kasungu District Socio-economic Profile*, KDPD, Kasungu.

Lenton, R. and Muller, M. (eds) (2009) *Integrated Water Resources Management in Practice: Better Water Management for Development*, Earthscan, London.

MA (Millennium Ecosystem Assessment) (2005) *Ecosystems and Human Well-being: Wetlands and Water Synthesis*, World Resources Institute, Washington, DC.

McCartney, M. P. (2000) 'The water budget of a headwater catchment containing a dambo', *Physics and Chemistry of the Earth*, vol. 25, nos. 7–8, pp. 581–591.

McCartney, M., Masiyandima, M. and Houghton-Carr, H. (2005) *Working Wetlands: Classifying Wetland Potential for Agriculture*, International Water Management Institute, Africa Regional Office, Pretoria.

McFarlane, M. J. and Whitlow, R. (1990) 'Key factors affecting the initiation and progress of gulleying in dambos in parts of Zimbabwe and Malawi', *Land Degradation and Rehabilitation*, vol. 2, no. 3, pp. 215–235.

Mortimore, M. J. and Adams, W. M. (2001) 'Farmer adaptation, change and crisis in the Sahel', *Global Environmental Change*, vol. 11, no. 1, pp. 49–57.

Msukwa, C. (2007) *Baseline PRA Report for Wetland Demonstration Sites, Kasungu District, Malawi*, Striking a Balance Project, Lilongwe.

Ndiyoi, M., Sampa, J. B., Thawe, P. and Wood, A. P. (2009) 'Striking a balance: maintaining seasonal dambo wetlands in Malawi and Zambia', in P. Mundy (ed.) *Planting Trees to Eat Fish: Field Experiences in Wetlands and Poverty Reduction*, Wetlands International, Wageningen, pp. 27–39.

Newson, M. (1997) *Land, Water and Development: Sustainable Management of River Basin Systems*, Routledge, London.

Ramsar Convention Secretariat (2010) *River Basin Management: Integrating Wetland Conservation and Wise Use into River Basin Management*, Ramsar handbooks for the wise use of wetlands, 4th edition, vol. 9, Ramsar Convention Secretariat, Gland.

Raymond, C. M., Fazey, I., Reed, M. S., Stringer, L. C., Robinson, G. M. and Evely, A. C. (2010) 'Integrating local and scientific knowledge for environmental management,' *Journal of Environmental Management*, vol. 91, no. 8, pp. 1766–1777.

Roberts, N. (1988) 'Dambos in development: management of a fragile ecological resource', *Journal of Biogeography*, vol. 15, no. 1, pp. 141–148.

Roberts, N. and Lambert, R. (1990) 'Degradation of dambo soils and peasant agriculture in Zimbabwe', in J. Boardman, I. D. L. Foster and J. A. Dearing (eds) *Soil Erosion on Agricultural Land*, John Wiley, London.

Sampa, J. (2008) *Sustainable Cultivation of 'Acid' Dambos*, Striking a Balance Project, Mpika, Zambia.

Sampa, J., Thawe, P., Kafuwa, D. and Dixon, A. (2008) *Wetland Institutions and Sustainable Management of Natural Resources: Experiences in Zambia and Malawi*, Striking a Balance Project, Simlemba and Mpika.

Sandbrook, C. (2012) *Biodiversity, Ecosystem Services and Poverty Alleviation: What Constitutes Good Evidence? A Discussion Paper*, Biodiversity, Ecosystem Services and Poverty Alleviation: Assessing the Current State of the Evidence, an ESPA Evidence and Impact Research Grant Project implemented by IIED and UNEP-WCMC.

Shaxson, T. F. (2006) 'Re-thinking the conservation of carbon, water and soil: a different perspective', *Agronomie*, vol. 26, no. 1, pp. 1–9.

Thawe, P. (2008) *Function Landscape Approach to Sustainable Wetland Management: Integrating Wetland and Catchment Management in Simlemba TA, Kasungu District, Malawi*, Striking a Balance Project, Simlemba, Malawi.

Tiffen, M., Mortimore, M. and Gichuki, F. (1994) *More People, Less Erosion: Environmental Recovery in Kenya*, John Wiley, Chichester.

Trapnell, C. G. and Clothier, J. N. (1937) *The Soils, Vegetation and Agriculture Systems of North-Western Rhodesia*, Government of Northern Rhodesia, Lusaka.

Verhoeven, J. T. A., Soons, M. B., Janssen, R. and Omtzigt, N. (2008) 'An operational landscape unit approach for identifying key landscape connections in wetland restoration', *Journal of Applied Ecology*, vol. 45, no. 5, pp. 1496–1503.

von der Hyden, C. J. and New, M. G. (2003) 'The role of a dambo in the hydrology of a catchment and the river network downstream', *Hydrology and Earth System Sciences*, vol. 7, no. 3, pp. 339–357.

Wood, A. P. (2005) *Sustainable Wetland Management for Livelihood Security, Simlemba TA, Kasungu District, Malawi, an Environmental and Socio-Economic Impact and Development Assessment*, Wetland Action, Huddersfield.

Wood, A. P. (2011) *GAWI Analysis of Seasonal Wetlands in Malawi: Towards Sustainable Multiple Use*, Wetland Action, Huddersfield.

Wood, A. P. and Dixon, A. B. (2009) 'Wetland use as an adaptive response to climate change in east and southern Africa: sustainable wetland management and the functional landscape approach', *Policy Briefing Note 5*, Striking a Balance, Simlemba and Mpika.

Wood, A. P. and van Halsema, G. (eds) (2008) *Scoping Agriculture-wetland Interactions: Towards a Sustainable Multiple Response Strategy*, FAO, Rome (Water Report 33).

Woodhouse, P., Bernstein, H. and Hulme, D. (2000) *African Enclosures? The Social Dynamics of Wetlands in Drylands*, James Currey, Oxford.

4 Local institutions, social capital and sustainable wetland management

Experiences from western Ethiopia

Alan Dixon, Afework Hailu and Tilahun Semu

Summary

Drawing on field research carried out in western Ethiopia over the last 15 years, this chapter examines the potential contribution that local people can make towards sustaining wetland ecosystem services (ESS) and livelihood benefits through the use of traditional, local institutional arrangements. As reported elsewhere in the literature, such local institutional arrangements that evolve from local peoples' knowledge and understanding of their immediate economic, social and physical environment, and that embody a form of social capital and shared management goals, have increasingly been regarded as effective mechanisms for the sustainable management of common property resources (CPRs) in developing countries. The chapter suggests that whilst local institutions are common throughout western Ethiopia, and do play a critical role in wetland management, contrary to the literature many are unable to function effectively today without the support of external actors such as government, courts or non-governmental organizations (NGOs). Although this could be interpreted as indicative of weak social capital and failing local institutions, we argue that locally driven requests for external support in the development of local institutional arrangements provide a clear mandate and opportunity for closer engagement between policy-makers and local people in the wetland management process.

Local institutions for natural resource management

Within the global discourse of wetland management there has arguably been an increasing (somewhat belated) recognition that local people are wetland managers as well as wetland resource users (Chapter 1). This has been influenced by the wider natural resource management (NRM) and development literature over the last 30 years, much of which has rejected simplistic neo-Malthusian notions that people automatically exploit and degrade their environment, instead drawing attention to a wealth of empirical research that highlights the complex, dynamic and adaptive nature of people–environment interactions. During the late 1970s and early 1980s in particular, the idea that the poor and marginalized in the developing world could be knowledgeable about their natural resources and even capable of managing them in a sustainable manner, began to be influential in mainstream development thinking (Howes and Chambers, 1979; Swift, 1979; Brokensha *et al.*, 1980; Chambers, 1983; Richards, 1985). Brokensha *et al.* (1980), for example, were among the first to highlight how local people drew on detailed 'indigenous knowledge' (IK) of their environment to organize NRM activities. Subsequent research throughout the 1980s and 1990s went on to explore the dynamic and adaptive nature of IK and, critically, how this underpins adaptive and sustainable community-based natural resource management (CBNRM) (Chambers *et al.*, 1989; Tiffen *et al.*, 1994; Warren *et al.*, 1995; Leach and Mearns, 1996; Reij and Waters-Bayer, 2001; Folke *et al.*, 2002).

During the late 1990s the Department for International Development's (DFID) Sustainable Livelihoods Framework (SLF) further consolidated the importance of local people and their knowledge by embedding IK within the 'livelihood assets' of human and social capital (see Chapter 1); the tangible and intangible resources that influence peoples' capacity to pursue and develop a livelihood (Carney, 1998; Scoones, 1998; Ellis, 2000). In recognizing 'social capital' as a livelihood asset the SLF acknowledges that the acquisition, evolution and application of IK is dependent upon community-based relations of trust and reciprocity, common rules and values, connectedness and knowledge networks (Putnam, 1995; Grootaert, 1998; Pretty and Ward, 2001). Popularized as the 'missing link in the development equation' (Mazzucato and Niemeijer, 2002) social capital has been regarded by many as playing a central role in empowering IK, social cohesion, local adaptive capacity and resilience, and ultimately sustainable NRM within socio-ecological systems (SES) (Grootaert, 1998; Ostrom, 2000; Pretty and Ward, 2001; Folke *et al.*, 2002; Adger, 2003; Pelling and High, 2005). Its critics, meanwhile, point to its nebulous nature, problems of measurement, and the fact that it can also embody and entrench inequitable power relations and hence hinder development (see Quibria, 2003).

In the context of CBNRM, social capital manifests itself in one way as community-based organizations (CBOs) and local institutions that organize and facilitate collective action through negotiated rules, norms and patterns of behaviour (Mearns, 1995; Manig, 1999; Mazzucato and Agrawal, 2001; McAslan, 2002; Niemeijer, 2002; Dasgupta, 2003). Local institutional arrangements include shared formal or informal rules and understandings that can relate to issues such as resource boundaries, land access and distribution, collective action, reciprocal work arrangements and conflict resolution (Rasmussen and Meinzen-Dick, 1995). Being endogenous in nature they have been regarded as 'home grown' adaptations to NRM problems (Shivakumar, 2003), and hence have greater credibility, relevance and legitimacy for local communities, compared to top-down, externally driven institutions. In terms of sustainability, empirical evidence suggests that they represent socially sustainable structures that support environmentally and economically sustainable resource management (Ostrom, 1990; Uphoff, 1992; Blunt and Warren, 1996; Howes, 1997; Hinchcliffe *et al.*, 1999; Koku and Gustafsson, 2001; Pretty and Ward, 2001; Mazzucato and Niemeijer, 2002). It is perhaps unsurprising, therefore, that NGOs and government agencies have increasingly sought to capitalize on this relationship and place the empowerment of local institutions on the development agenda (Ribot *et al.*, 2008).

One key area of debate, however, centres on the enabling/facilitating conditions required for community-based local institutions to function effectively and facilitate the sustainable management of natural resources (see Box 4.1; Ostrom, 1990; Agrawal, 2001; Cox *et al.*, 2010) and, concomitantly, the extent to which these institutions can remain socially sustainable in the face of intervention from external institutions such as NGOs or government (Ostrom, 2000; Agrawal, 2001). Of primary concern is that external intervention in the operations of local institutions constitutes a threat to their legitimacy, credibility and effectiveness (Richards, 1997; Serra, 2001; Watson, 2003). An examination of the CPR management literature, however, reveals a more complex situation; 'supportive external sanctioning institutions' are cited as one of the 'enabling conditions for the sustainability of the commons' by Agrawal (2001: 1659), as is 'Central government should not undermine local authority'. Similarly, Rasmussen and Meinzen-Dick (1995) cite the work of Wade (1988), Bardhan (1993) and Ostrom (1990) in arguing that local institutions become more effective when arrangements in the external environment support them. As illustrated in Box 4.1, for Ostrom (1990: 90), ensuring that 'the rights of appropriators to devise their own institutions are not challenged by external governmental authorities', is one of eight fundamental design principles that characterize the efficacy of local institutions in facilitating sustainable CPR management. Clearly, the implication here is that it is not the intervention per se but the nature of it that is the critical variable; any intervention must support and respect the rights and decision-making power of the local institution.

Despite the interest in local institutions among development practitioners, mainstream wetland management policy-makers have thus far failed to engage with these ideas in any significant way. Despite recent shifts in the wetland management agenda towards livelihoods and poverty reduction (see Chapter 1), there remains significant scope for devolving wetland management decision making to community-based local institutions. Indeed, decades of development literature suggests that without the full participation or engagement of local communities, any conservation or development initiatives are unlikely to be sustainable or successful (Chambers, 1983; Craig and Mayo, 1995; Pretty, 1995; Fisher *et al.*, 2005).

Box 4.1 Elinor Ostrom's design principles for long-enduring CPR institutions

Source: Ostrom, 1990: 90

1. *Clearly-defined boundaries*
   Individuals or households who have rights to withdraw resource units from the CPR must be clearly defined, as must the boundaries of the CPR itself.

2. *Congruence between appropriation and provision rules and local conditions*
   Appropriation rules restricting time, place, technology and/or quantity of resource units are related to local conditions and to provision rules requiring labour, material and/or money.

3. *Collective choice arrangements*
   Most individuals affected by operational rules can participate in modifying the operational rules.

4. *Monitoring*
   Monitors, who actively audit CPR conditions and appropriate behaviour, are accountable to the appropriators or are the appropriators.

5. *Graduated sanctions*
   Appropriators who violate operational rules are likely to be assessed graduated sanctions (depending on the seriousness and context of the offence) by other appropriators, by officials accountable to these appropriators or by both.

6. *Conflict resolution mechanisms*
   Appropriators and their officials have rapid access to low-cost arenas to resolve conflicts among appropriators or between appropriators and officials.

7. *Minimal recognition of rights to organize*
   The rights of appropriators to devise their own institutions are not challenged by external government authorities.

   For CPRs that are parts of larger systems:

8. *Nested enterprises*
   Appropriation, provision, monitoring, enforcement, conflict resolution and governance activities are organized in multiple layers of nested enterprises.

This chapter now considers these ideas in the context of a case study from western Ethiopia, where research carried out over the last 15 years has drawn attention to the importance of local people and their institutions in facilitating the maintenance of multiple ESS from wetlands. In particular, the discussion examines the functioning and effectiveness of local wetland management institutions (LWMIs), and the extent to which both have been influenced and facilitated by relationships with external actors. In doing so, it seeks to elicit transferable lessons that can inform the future engagement of wetland policy-makers and other actors with local institutions.

Wetlands in western Ethiopia

Situated in East Africa, Ethiopia is a geographically and culturally diverse country. Its landscape ranges from lowland desert in the Afar Region (125 m below sea level) to dissected highland plateau, tropical montane rainforest and some of the highest mountains in Africa (over 4,000 m above sea level). The interaction between topography and climate create significant spatial diversity in ecosystems, and riverine, lacustrine and palustrine wetland landforms together account for approximately 1.14 per cent of the total land area (Hillman and Abebe, 1993; Abebe, 2003).

The western highlands, whose topography ranges from 1,000 to 2,000 m, are characterized by rainfall that is frequently in excess of 1,500 mm, and an average temperature of around 20 °C. This gives rise to a warm moist temperate climate atypical of conditions in the rest of country. It is also one of the most fertile and least exploited parts of the country, and although deforestation has increased in recent years, significant areas of tropical montane rainforest (*Aningeria adolfi-friederici, Croton macrostachyus* and *Sapium ellipticum*) remain (Gole, 2003). The highland plateau is dissected by numerous flat-bottomed, poorly drained river valleys that facilitate the formation of both permanent and seasonal wetlands ranging in size from less than 10 ha to more than 300 ha. Of these, small headwater wetlands are more abundant particularly in the Illubabor and Western Wellega zones (Figure 4.1) (Dixon, 2003), where wetlands account for approximately 4 per cent of the total land area (Hailu *et al.*, 2000b). Throughout the western highlands, wetlands play an important role in the everyday lives of poor rural communities through the provision of numerous ESS (Table 4.1). Peripheral springs are a source of clean drinking water throughout the year for local people, and the natural wetland sedge, known locally as *cheffe* (*Cyperus latifolius*), is used for thatching, handicrafts and religious ceremonies (Hailu *et al.*, 2000b). A range of medicinal plants are also collected by those living around the wetlands (Woldu, 2000).

These wetlands are also valuable agricultural resources as the availability of moisture supports the cultivation of maize and other crops during the dry periods of the year (January to April); wetland crops are subsequently harvested at a time when stores of upland food crops cultivated during the

Figure 4.1 Western Wellega and Illubabor Zones within Ethiopia

preceding wet season (June to September) are depleted – the so called 'hungry season'. Figure 4.2 illustrates a typical agricultural calendar for wetland farmers, during which the main period of wetland cultivation is characterized by the preparation and management of drainage ditches, the burning of crop residues, ploughing, sowing, weeding and harvesting. Throughout the whole period farmers are continuously engaged in the monitoring and management of soil moisture levels through either ditch excavation or blocking, and following sowing, the guarding of crops against wild pests and cattle damage. Traditionally, wetland cultivation was abandoned after the harvest in July, and ditches were blocked to allow water levels to recover and natural vegetation to regenerate. In line with Ethiopian law, the wetlands themselves remain the property of the state but individuals possess hereditary 'lifelong usufruct rights' to plots within wetlands (see Crewett *et al.*, 2008).

Table 4.1 Wetland ESS and beneficiaries in the Illubabor Zone

Provisioning services	Estimate of households benefiting
Cultivation	25%
Water for stock	Most cattle owners (30% of population)
Dry season grazing	Most cattle owners (30% of population)
Domestic water from springs	50–100%
Craft materials (palm products and sedges)	5%
Medicinal plants	100% (mostly indirectly by purchase from collectors/traditional doctors)
Thatching reeds	85% (most rural households)
Temporary crop of sedges guarding huts	30%
Regulating services	Likelihood of occurrence based on MA (2005)
Climate regulation	Low
Hydrological regimes	Medium
Pollution control	Low
Flood control	High
Cultural services	
Social/ceremonial use of sedges	High (most households)
Support services	
Biodiversity	Medium (see Woldu, 2000)
Soil formation (sediment retention)	Medium
Nutrient cycling	High
Pollination	Low

Sourcce: adapted from Hailu *et al.*, 2000b

Although the precise origins of wetland farming remain unclear, anecdotal evidence suggests that this practice probably emerged in the area during the late nineteenth century via the dissemination of wetland farming knowledge along trade routes from the north and east (McCann, 1995; Dixon, 2005). What is clear, however, is that the agricultural use of wetlands has increased in recent decades in response to the demand for food driven by numerous factors including population pressure, the commercialization

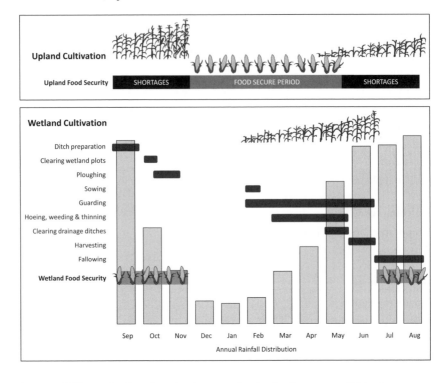

Figure 4.2 The upland and wetland seasonal calendar in western Ethiopia

of agriculture, the expansion of coffee production in the uplands (which reduces farmland availability), government policies that set regional food security targets (Asres, 1996; Dixon, 2003) and, more recently, the need for farmers to retain security of tenure in the face of government threats of land expropriation. Although during the first half of the twentieth century, wetlands (and individual plots within wetlands) were cultivated only when required to address acute food shortages, more and more wetland areas have been permanently cultivated since the 1960s. In many cases intensive drainage regimes have been implemented that allow cultivation during the wet season, in addition to the dry season, so that two crops per year are produced. Estimates suggest that wetland cultivation provides between 10 per cent and 20 per cent of the annual food needs of Illubabor and Western Wellega's population (Ethiopia Network on Food Security, 2001), but during the food shortage months, its contribution is up to 100 per cent in some areas. In Western Wellega, the dependence on wetlands for food security is higher than that in neighbouring Illubabor, on account of higher population densities (140.3 and 94.5 pers/km^2 respectively; Central Statistics Agency, 2010) and more extensively degraded uplands.

As wetland agriculture has become more extensive and intensive in the western highlands, concerns have been raised over the sustainability of their ESS. During the early 1990s reports from the field suggested that in many areas wetland cultivation was being abandoned as a result of falling water tables and declining soil fertility (Tato, 1993; Wood, 1996; Hailu *et al.*, 2000a). In simple ecological terms, it appeared that wetlands were being exploited beyond their carrying capacity, and there was concern that overexploitation would lead to the permanent loss of the provisioning, regulating, cultural and supporting services associated with these wetlands. In terms of rural livelihoods, this situation represents a loss of livelihood assets (natural capital) and hence a reduced likelihood of people achieving sustainable livelihood outcomes such as a regular income, food security and reduced vulnerability to change. The continuing demand for food, driven in part by government food production campaigns, meant that despite these indications of degradation occurring, local people had little choice but to carry on using wetlands.

Researching human and social capital

As a response to these emerging issues, a collaborative research project known as the Ethiopian Wetlands Research Programme (EWRP) was established in 1997, with the aims of investigating the environmental and socio-economic dynamics of wetland utilisation, and identifying ways of achieving sustainable wetland use in the Illubabor Zone. This project, which brought together researchers from the physical and social sciences in an interdisciplinary study, was the start of a range of research and implementation activities in the western highlands which, rather than seeking the imposition of draconian restrictions on wetland utilisation, sought first to understand the social, environmental and economic context of people–wetland relationships in the area, principally through engaging with local people and seeking to learn from their experiences.

The major findings of the EWRP research (see Wood and Dixon (2002) for a summary of the programme's outcomes) identified two significant features of wetland management in Illubabor. First, contrary to initial expectations and the original rationale for the project, there appeared to be little evidence of long-term wetland degradation in those sites studied. Although the research drew attention to the ecological, hydrological and pedological changes taking place as a result of agricultural drainage, evidence suggested that these changes were reversible and the regeneration of 'normal' wetland conditions could occur over a period of a few years, typically three to six, once wetland drainage ceased (Dixon, 2003). Second, and perhaps more significant, was the recognition that wetland management was not carried out in a disorganized, *ad hoc* manner by local people. On the contrary, participatory investigations revealed that wetland management strategies, including coordination of drainage and agricultural activities, were rooted in local people's detailed knowledge of the dynamic wetland environment, and an understanding of the effects of their utilization on this environment. Local people also had an intimate understanding of the mechanisms of environmental sustainability in the wetlands, in terms of the processes of degradation and regeneration and how these related to ecohydrological changes. Moreover, in some (but not all) cases, community-based wetland management coordinating committees ensured that wetland management was carried out in an efficient, environmentally sensitive and sustainable manner. Human and social capital, using the terminology of the SLF (Chapter 1: Figure 1.3), was evidently playing a key role in sustaining the benefits from wetlands.

Subsequent investigations carried out in Ilubabor and Western Wellega during the 2000s have shed further light on the role of human and social capital in these wetland management systems, and specifically the links with livelihood and environmental sustainability. With regards to human capital, although IK does play a significant role in determining the ways in which wetlands are used, it is arguably only as effective as the extent to which it can adapt and evolve in response to socio-economic and environmental changes (this relates to the SLF's 'vulnerability context' and the wider literature on adaptive capacity and resilience (see for example, Folke *et al.*, 2002). As the terms of engagement with wetlands change for local people, e.g. as a result of climate change, population pressure or new government food security policies, new knowledge needs to be acquired in order for wetland management practices to adapt and remain sustainable. In Illubabor, participatory research exploring wetland knowledge acquisition and development suggested that innovation and experimentation among wetland farmers, in terms of crops and water management, was common and widespread, and led to the generation of new wetland knowledge. This in turn facilitated the adaptation and evolution of management practices characterized by, for example, new cropping regimes and water conservation measures, which appeared to have a positive impact on environmental and livelihood outcomes (Dixon, 2005). As well as citing numerous examples of innovation, wetland users also recalled how wetland knowledge had been acquired over the years through a range of indigenous and external communication channels ranging from talking to other wetland users at local markets and personal observation (curiosity), to selectively adopting and incorporating information from ancestors and development agents (Dixon, 2005). Moreover, the social context of community and the wetland management co-ordinating committee were cited as key sources of information and mechanisms of knowledge exchange. In those areas where extensive communication networks and innovation were evident, there were indications that a range of ESS was being sustained, yet in approximately 10 per cent of cases where communication between wetland users was poor, and where experimentation seldom occurred, wetlands exhibited signs of degradation (e.g. low water table, soil compaction and mineralisation), were unusable and hence subsequently abandoned (Dixon, 2005).

Social capital, therefore, in terms of knowledge networks, common wetland values and relationships of reciprocity, appears to have played a critical role in facilitating the acquisition and development of wetland knowledge in the area. But it is perhaps social capital as manifest in local institutional arrangements for managing wetlands that has arguably had the greatest influence on the sustainability of wetlands and livelihoods.

Social capital as local wetland management institutions

While the original EWRP study identified one example of an LWMI in its study areas, subsequent research has shown that these institutions are common throughout the western highlands, due not least to the nature of wetland agriculture itself which requires significant coordination in terms of labour and resources. Although there are some spatial and temporal variations in their structure and functioning (see Dixon, 2008), LWMI membership typically includes all those who have usufructary rights to plots of land within wetlands. The administrative committee of each LWMI usually consists of one elected leader and treasurer who report upwards to the kebele (sub-district) administration, and who are accountable to all the wetland farmers (Figure 4.3). The main aim of the LWMIs is to coordinate wetland management activities, mainly to support crop production but also to sustain the resource base of the wetlands themselves. Typically, they will institute arrangements so that drainage, ploughing, sowing and ditch-blocking are undertaken by all users at the same time (Table 4.2), and these are often enshrined in a written constitution. One increasingly important task is ensuring all users carry out pest-guarding duties during the cultivation period to reduce the risk of crop loss caused by baboons and pigs (Quirin and Dixon, 2012). Reciprocal working arrangements are also common; wetland users may provide labour for each other during arduous tasks such as drainage, and also loan out draught animals during ploughing.

Whilst the majority of the rules and functions of the LWMIs relate to crop production, some specifically deal with issues of environmental sustainability, and hence demonstrate wetland users' knowledge of the interrelationships between different ESS. For example, LWMIs make informed decisions on whether whole wetlands should be cultivated, reserved for *cheffe* production or if the wetland is perceived as being degraded, abandoned and left to regenerate. They have regulations relating to the 'correct' depth and width of drains, their layout and the reservation of plots of natural *cheffe* vegetation within wetlands in order to avoid over-drainage and crop failures, as well as wetland degradation. Similarly, some LWMIs prohibit the grazing of cattle in wetlands to avoid soil compaction, while the cultivation of crops, such as tef, sugar cane and even eucalyptus, are prohibited due to their capacity to reduce soil moisture content. This is illustrated in the words of one farmer:

Some people like to plant potato and tef after the maize is harvested, but so far no one has planted eucalyptus trees on the wetland we are cultivating. If someone wants to, the committee will stop it. Usually following the maize harvest the wetland is fenced, drainage ditches are blocked and the land is allowed to flood.

Farmer at Hadesa Wetland, Illubabor (2003)

As suggested here, if a wetland user fails to carry out their assigned activity or fails to adhere to the rules of the LWMI, then the committee reserves the right to impose a penalty, usually a fine, or refer individuals to the kebele court. This seldom occurs, however, since most wetland users appear to benefit – through improved crop yields – from their participation in the

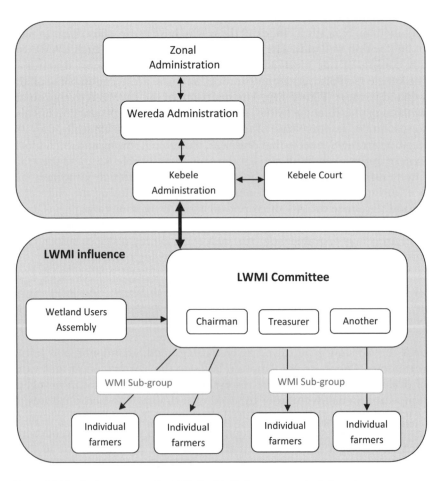

Figure 4.3 Typical structure of a LWMI with links to state administrative structures

Table 4.2 Commonly cited rules of the wetland management institutions and their functions

Constitutional rules/functions	Rationale/impacts
Co-ordination of ditch preparation	Ensures that all wetland farmers prepare their ditches at the same time to facilitate optimum flow of water for crop production
Construction of paths and bridges in wetlands	Bridges allow access within wetlands during wet season
Co-ordinating forest protection	Prevents indiscriminate upland deforestation
Co-ordination of the correct timing for hoeing and weeding	Minimizes the impact of weeds spreading between plots, hence improved moisture and fertility
Conflict resolution	Settles disputes over plot boundaries, negligent guarding and breaking the LWMI rules
Co-ordination of ploughing and sowing	'if someone lags behind then the whole [wetland] system will suffer' (Somie Wetland Farmer, Illubabor, 2003)
Co-ordinating the equitable distribution of water	Ensuring transfer of water from plot to plot
Guarding against wild pests	Committee organizes farmers into sub-groups who arrange 24-hour guarding schedules
Prevent wetland clearing/retain *cheffe*	Fallowing allows regeneration of natural vegetation, soil moisture and fertility
Prevent destructive practices (tef, sugar cane, eucalyptus)	Prevents some farmers cultivating the cereal crop tef (*Eragrostis tef*) immediately after the maize harvest. Tef cultivation prevents the recovery of the water table during the wet season and is widely considered as damaging to wetland health. Some crops are also recognized as water-demanding, hence damaging
Reporting to kebele	Those who ignore the rules are reported to the kebele administration and required by threat of fine or imprisonment to respect them

institutional arrangements and so they adhere to the guidance given. The only issue arising on a regular basis tends to be the neglect of pest-guarding duties leading to the destruction of crops; under the constitution of the LMWI, those whose crops are damaged should be reimbursed by those who

neglected their duties, although in most cases a more pragmatic concilia-
tory approach is taken.

The question arises: do these institutions and the apparent social capital
on which they are based help achieve sustainable use and, therefore, create a win–win situation for wetlands and livelihoods? Despite their focus
on co-ordinating actions that seek to achieve a balance in the provision of
wetland services, no research to date has been able to conclusively prove
a link between tangible environmental and livelihood outcomes and the
presence of these LWMIs. Nonetheless, there is little doubt that the presence of functioning LWMIs in wetland communities significantly reduces
the likelihood of poor communication and disagreements occurring
between wetland stakeholders, the proliferation of destructive management practices and the imposition of inappropriate, top-down management
strategies, all of which have been recognized from previous research as
key drivers of unsustainable use (Wood, 1996; Dixon, 2005). Moreover,
the LWMI arrangements appear congruous with Ostrom's prerequisites
for successful CPR institutions (Box 4.1), not least in terms of the clearly
defined resource boundaries involved, opportunities for resource user participation and the inherent conflict resolution mechanisms, sanctions and
monitoring systems. At the very least, LWMIs appear to mitigate the impacts
of drainage and cultivation on the wetland environment; at the interface
of society and the environment they constitute the adaptive space in which
local communities reconcile external pressures with their environmental
knowledge of wetlands.

Creating local institutional arrangements for managing wetlands

It has been argued that there are real development and environmental
advantages to be had in empowering social capital and supporting such
institutional arrangements (Warren *et al.*, 1995; Blunt and Warren, 1996;
Ostrom, 2000; Pretty and Ward, 2001). This is certainly the case in areas
where wetlands and their ESS are at risk of degradation, or where degradation is already occurring. As highlighted earlier in this chapter, any wetland
policy interventions that build upon local knowledge and empower LWMIs
from the bottom-up are more likely to be sensitive to local environmental
conditions, locally credible and sustainable themselves, as well as producing sustainable results in environmental, social and economic terms.
However, the empirical literature also raises concerns over the effectiveness
of local institutions that have been driven or influenced by external agencies, regardless of how sensitively this is done; local institutions are deemed
effective because they are local, and any attempt by outsiders to manipulate
community structures or in effect 'create' social capital, could be doomed
from the start (Richards, 1997; Serra, 2001; Watson, 2003). This raises
important questions for wetland managers and policy-makers in terms of

their ability to create and support 'local' institutional arrangements that are as economically, environmentally and socially sustainable as traditional, indigenous structures.

Closer examination of the situation in western Ethiopia, however, reveals that the relationship between LWMIs and external actors has always been complex and dynamic. Although it was initially assumed that all LWMIs in the area were traditional in nature, reflecting the indigenous Oromo *Abba Laga* community-based institutions of the past (Hassen, 1990), recent research undertaken in the area has highlighted the existence of more recent LWMIs whose origins are linked to the policies of the Derg (military) government era (1974–1991) when cooperative agriculture was actively encouraged. The reality throughout much of the area is one of externally-driven local institutions existing alongside more traditional forms (Box 4.2).

Box 4.2 Two experiences of wetland management institutions in western Ethiopia

a) *Shenkora – an externally driven LWMI*

Shenkora Wetland, located in Metu Wereda (district), Illubabor Zone, has over the years provided water, reeds, medicinal plants, grazing and agricultural land for local farmers. The cultivation of the wetland fringes is said to have originated during the Haile Selassie period (1930–1974), and this was later extended in 1982 on the request of the wereda (district) Ministry of Agriculture (MoA) during the Derg regime. According to farmers:

> A general instruction was passed to all kebeles to drain and cultivate wetlands, and they suggested that we organize ourselves to cultivate in a coordinated manner. Then we discussed it among ourselves, started draining the wetland and established the committee.
>
> Farmer at Shenkora Wetland, Illubabor (2003)

Farmers formed a wetland management committee initially composed of 24 farmers led by three team leaders, each responsible for co-ordinating wetland cultivation, grazing, crop guarding and reporting any breaches of the rules to the kebele administration:

> the kebele administration would act immediately if any problem was reported to them by the committee, and they took the necessary measures. The penalty ranged from 3 to 15 birr or imprisonment for up to five days. We had written rules and regulations signed by all wetland users ... everyone strictly did what they were told.
>
> Farmer at Shenkora Wetland, Illubabor (2003)

This situation appears to have continued up until the change in government in 1991, when a policy of redistributing wetland plots to all kebele members was implemented and farmers were instructed to reorganize into one large committee. Redistribution led to smaller wetland plot sizes for most farmers (less than 0.25 ha) and the allocation of plots to those with little interest in wetland management:

> Wetland was given to those who don't cultivate their wetland properly and they created a problem for us. They don't drain their wetland and the water from their wetland has an impact on our plots. There are so many problems. Some of these farmers are also members of the kebele administration. The kebele administrators are young people. They are not mature people. They expect others to work for them. But no one is willing to work for them. Thanks to the current democracy ... [laughter]
> Farmer at Shenkora Wetland, Illubabor (2003)

The emerging consensus among farmers at Shenkora Wetland is that the livelihood benefits derived from wetlands have steadily decreased since land redistribution and the change in government, mainly due to a wetland management committee that has been weakened by an unsupportive external environment. In the words of one farmer:

> In reality there is not much difference between the responsibility of the two committees operated under the two governments, but the major difference is that the current committee is not able to fulfil its responsibility as much as those operating under the Derg. They don't have the authority to take measures on those who fail to cultivate. Even if the committee reports the issue to the kebele administration, they won't do anything. They are very reluctant to support the wetland management committee.
> Farmer at Shenkora Wetland, Illubabor (2003)

b) *Laga Wajoti – a traditional LWMI*

According to local farmers, Laga Wajoti wetland in Gimbi Wereda, Western Wellega, has been used for water supply, grazing, reed production, medicinal plant collection and cultivation for well over 100 years, dating back to the Oromo administration (pre-1870s). During this period, a local institutional arrangement known as *Abba Koto* was responsible for managing both upland and wetland use, although this appears to have been the name ascribed to one person within the community who retained the power to make decisions over all NRM. With the shift to a feudal-style system of land ownership during the Haile Selassie period, in which farmers worked for landlords, *Abba Koto* was renamed *Abba Laga* and given responsibility solely for overseeing wetland management. In return for these duties, *Abba Laga* was gifted approximately one-tenth of the land he coordinated.

It was not until the land reform of the Derg in 1974, however, that the *Abba Laga* became freely elected by, and amongst, wetland users themselves. Since 1991 a further two 'members' have been added to the *Abba Laga* (throughout its history, the term *Abba Laga* has been used interchangeably to describe both the institution itself that is made up of participating farmers, *and* the appointed head of the institution). Interestingly, this decision to increase the management committee of *Abba Laga* came from the farmers themselves, in response to concerns about disorganized wetland cultivation. This, however, does not appear to stem from a poor relationship with the kebele administration, as was the case in Shenkora Wetland (above). On the contrary, it was suggested that:

> We have a strong relationship. The kebele administration acts automatically on the problems we report to them. That is why our *Abba Laga* functions better than others. Although we have a number of problems to be solved, we think we are doing better than neighbouring *Abba Lagas* in our kebele.
>
> They [kebele administration] know that wetlands are the most important natural resource that most people depend on for food supply here in our catchment. If they don't pay proper attention to wetlands the problem of food shortage will be so serious that they can't solve it. That is why we think they have paid serious attention. They are also concerned with their kebele's development and creation of better welfare for the community.
> Farmer at Laga Wajoti Wetland, Western Wellega (2003)

Indeed, farmers cite the lack of a written constitution that outlines the rules and regulations of wetland use, as the main threat to effective wetland co-ordination:

> We have rules and regulations relating to things like preventing livestock coming into wetlands for grazing, but this is not enough. We have to have more detailed rules and regulations, including details of penalties.
>
> If there are no abiding rules and regulations, members become reluctant to take instructions from *Abba Laga* and implement them ... A copy has to be submitted to the kebele administration.
> Farmer at Laga Wajoti Wetland, Western Wellega (2003)

What is clear from the case studies in Box 4.2, which are typical of the range of LMWI structures found throughout the area (Dixon, 2008), is that LMWIs do not exist in an institutional vacuum. Even where wetland institutional arrangements have evolved directly from traditional institutions over a long period of time, it appears that their structure and functioning have been influenced by an enabling external political environment in the past, which has supported (or at least not sought to undermine) local institutional arrangements for wetland management. Indeed, in both of the cases highlighted above the current survival of the existing wetland institutional arrangements appears to hinge on the backstopping and political support provided by the government via the kebele administration. Yet somewhat paradoxically, this support to enforce the rules of the wetland institutions is needed because, as farmers suggest, the government itself has undermined the ethos of collective action by repeatedly espousing (especially during public meetings) its commitment to, and support of, democratic rights and individual freedoms. Those 'democratically empowered' farmers who subsequently opt-out of the LWMIs on the grounds that they want to manage their wetland plots on their own terms, are blamed for institutional weakening but also for increasing the workload for the remaining LWMI farmers; having uncultivated plots adjacent to those that are cultivated renders hydrological management significantly more challenging. Thus, in this complex, socio-political environment, the government simultaneously provides both the enabling and disabling conditions for these local wetland institutional arrangements. Both scenarios concur with the facilitating conditions for CPR institutions identified in Agrawal's meta-analysis of the literature (Agrawal, 2001), but critically the situation also seems to be one where, in Ostrom's terms, the government is increasingly challenging 'the rights of appropriators to devise their own institutions' (Ostrom, 1990: 90).

The Ethiopian experiences also raise some questions regarding the role of social capital *per se* in these institutions. Whilst there is evidence of shared goals and common values, trust and reciprocity among wetland-using communities, the fact that some farmers are beginning to ignore the rules of the institutions, is symptomatic of a breakdown in social capital and the ideals of collective management. Although farmers' response, in terms of seeking more engagement from government in the enforcement of rules, may have a positive effect on the functioning of LWMIs, such actions may only erode social capital further by fostering wetland management characterized by top-down coercion rather than bottom-up cooperation. Nonetheless, it has to be acknowledged that this is what farmers desire, and what they regard as being the answer to existing failures in their wetland management institutions, and in that sense it constitutes a form of bottom-up planning.

In terms of balancing environmental and development outcomes from wetlands, the issues that emerge from the Ethiopian experience confirm that in some cases external actors have a key role to play in facilitating the 'enabling' conditions in which LWMIs can function more effectively (and in this sense they offer a critique of the often romanticized view of 'indigenous' institutions managing natural resources in isolation). Intervention by NGOs, government and other external actors can, therefore, still be advantageous and desirable, providing that any policy-making or other intervention is carried out in a participatory, bottom-up manner that respects the knowledge, needs and aspirations of local people, and is supportive of the local institutional environment.

As suggested in Box 4.2 above, the situation in Ethiopia has been one where external policy-makers such as government or NGOs were seldom involved in wetland management at the local level; wetlands were regarded as either agricultural resources under the remit of the Ministry of Agriculture and Rural Development (MARD), or as sources of irrigation water under the Ministry of Water Resources. Indeed, a series of interviews held with local government development agents in Illubabor and Western Wellega in 2003 revealed that few were even aware of the existence of wetland management institutions. There are, however, indications that this situation is changing and that since 2008 MARD has been working closely with *Abba Laga* in some areas, albeit as a means of facilitating an increase in the agricultural output from wetlands. Clearly there are dangers here in terms of environmental sustainability, and it remains to be seen whether these local institutions can exert an influence, retain their decision-making and management rights, and mitigate external pressures that appear to be driven by the government's short-term economic concerns.

The NGO sector, meanwhile, has been more active in recognizing the linkages between local institutions and sustainable development, and since 2000 the Ethio-Wetlands and Natural Resources Association (EWNRA), Ethiopia's first wetlands-dedicated NGO, has been undertaking a range of research, advocacy and implementation activities with wetland farmers in several parts of the country. Building on the premise that local institutions can have a positive impact on the management of wetland ESS, the EWNRA implemented a project in 2005 entitled 'Integrated Wichi Wetland and Watershed Project' (see Box 4.3). This project focused on one large wetland of around 364 ha, known locally as Wichi, located in the centre of Illubabor Zone, and aimed to 'Improve the economic and environmental values of Wichi wetland and its surrounding watershed and thereby contribute towards food security and livelihood enhancement of the local communities'. The project was a direct response to reports from local farmers that environmental services associated with the wetland were in decline following several decades of drainage and cultivation, driven by a combination of population

---

Box 4.3 EWNRA's Integrated Wichi Wetland and Watershed Project

Overall aim

To improve the economic and environmental values of Wichi wetland and its surrounding watershed, and thereby contribute towards food security and livelihood enhancement of the local communities.

Objectives

1) Reduce the level of land degradation resulting from soil erosion, overgrazing and deforestation.
2) Reduce wetland and wetland resources degradation.
3) Improve land productivity in the watershed through implementing biophysical soil and water conservation, and compost preparation and application.
4) Increase livestock productivity through improving grazing management and on-farm forage development.
5) Improve income and livelihood of the community through diversification of livelihood opportunities (skills-building training, revolving fund supply, diversifying homestead productivity through honey, vegetable and fruit production).
6) Build capacity within the community on NRM.
7) Empower the beneficiary community through strengthening their traditional organization with watershed committees and byelaws.

---

pressure, upland food shortages and government directives encouraging wetland drainage. In a household survey carried out in 2005 the majority of farmers reported a 'drying out' of the wetlands, shortages of *cheffe* grass and a decline in crop productivity. Although local institutional arrangements for managing the wetland had reportedly existed in the past, these had become ineffective and only reciprocal work arrangements known locally as *dado* and *debo* were being invoked for wetland tasks such as ditch digging.

Whilst Objective 7 of the project (Box 4.3) sought to empower 'traditional organizations', more emphasis was placed on the creation of new local institutional arrangements linking the management of both catchment and wetland. This was achieved through a range of participatory planning exercises held with farmers from the Wichi watershed, which drew on their past experiences of local institutional arrangements for wetland management. This led to the development of new byelaws that set limits on the use of wetlands for agriculture, and committed members to livestock guarding to prevent damage to wetland crops, designating wetland areas for grazing, as well as implementing soil and water conservation initiatives such as vetiver grass terraces in the catchment. The success of catchment rehabilitation measures, which resulted in an upland increase in crop yields of 30 per cent, subsequently refocused people's livelihood needs away from the agricultural use of the wetland, and by 2009 wetland degradation had reportedly been reversed as water levels increased and *cheffe* grass became abundant once again.

Some of the key project interventions and outcomes are highlighted in Box 4.4, but in terms of building local institutional arrangements the intervention of the EWNRA appears to date to have had a positive effect overall on wetlands and livelihoods in the area (to the extent that the EWNRA's intervention has subsequently been cited as an example of best practice by the regional government). Farmers currently derive a range of non-agricultural ESS from their regenerated wetland, whilst benefiting economically from improved catchment-based livelihoods. But whilst the EWNRA's experience provides lessons in terms of engagement with wetland-using communities and the empowerment of local institutional arrangements, its intervention has essentially been characterized by a more traditional preservationist approach that has sought to minimize, and almostly completely halt, agriculture in the wetland. Although the EWNRA argues that this was the desired outcome of farmers, who have always preferred to reserve wetlands for *cheffe* production and livestock grazing rather than cultivation, it has, nonetheless, resulted in the enhancement of some wetland ESS, e.g. regulating services and the provision of freshwater, alongside the loss of others, e.g. provisioning services such as crop cultivation and grazing resources (due to flooding). Whilst this may suit the farmers in the Wichi watershed who are able to access land in both the uplands and wetlands, one has to question the transferability of this approach to those areas where farmers are heavily dependent on wetlands. Although the Wichi experience has not culminated in the sustainable, multiple use of the wetland, it has demonstrated that: a) these wetlands can be rehabilitated via community-based management interventions, and b) local institutions are dynamic and flexible, and can adapt to new situations as evidenced by the transfer of focus away from the wetland to the catchment. Should the livelihood need arise in the future, the Wichi community arguably possesses the institutional and social infrastructure necessary to re-engage in more intensive forms of wetland use in a sustainable manner.

Box 4.4 Key outcomes from the Wichi project

- Natural resource management: 946 km of soil and water conservation structures implemented in the catchment; 255,000 tree seedlings planted on degraded land, farm boundaries and as woodlots around homesteads; 233 farmers involved in composting training.
- Livelihood diversification and income generation: 34,100 fruit tree seedlings distributed to 1,400 people; 80 beehives distributed; 135 women involved in the formation of a micro-credit scheme.
- Clean water supply and sanitation: eight handpumps installed at the wetland edge (4,500 people benefiting from access to clean water); nine water and sanitation committees established (for pump maintenance); 1,290 people trained on sanitation and hygiene.
- Awareness raising, capacity building and community empowerment: over 450 community representatives from existing local institutions such as *Abba Laga*, *Shene* and *Idir*, were trained on natural resource (including wetlands) management, beekeeping, fruit and vegetable production, agroforestry and institutional management.
- Formation of a watershed management committee made up of community representatives.

Conclusions: wetlands and local institutions in Africa

Within the western Ethiopian context, LWMIs, whether formed by farmers themselves or by external agents, have clearly played a central role in balancing ESS and livelihood benefits from wetlands. Irrespective of debates concerning the extent to which they embody social capital, or indeed the manner in which they were formed, field research has suggested that LWMIs are underpinned by a strong desire among the majority of wetlands users to derive benefits from wetlands in ways that are environmentally, socially and economically sustainable. Moreover, this desire is evidently based on an in-depth and dynamic knowledge of wetland environments, management practices and their implications, which wetland users possess. Put simply, LWMIs benefit local people, their livelihoods and wetlands themselves. The key issue in western Ethiopia is whether they can continue to do so, and in this sense there is a question over their effectiveness and sustainability in the long term. Whilst there are indications that LWMI arrangements are breaking down in some areas, there is nonetheless a desire among farmers to seek opportunities to strengthen these institutions, even if it means involving external actors such as government and NGOs in their function-ing. This in itself is an encouraging indication of an on-going process of adaptation, but success will ultimately hinge on the extent to which exter-nal actors respect and support local-level institutional decision-making and management structures, rather than overriding them.

A number of themes emerging from the Ethiopian research have potential implications for wetland management elsewhere in Africa, and can contribute to the wider discourse on wetland-based livelihoods. First, the Ethiopian experience has drawn attention to the important role that community-based local institutions can, and should, play in wetland man-agement. Because they are dynamic, adaptive, based on a shared interest and an evolving knowledge base, they can develop wetland management strategies that balance livelihood needs with ESS, and mitigate destruc-tive practices. As such, they arguably constitute the only legitimate means through which interventions for wetland-based livelihoods and sustainable wetland management can be coordinated and achieved. This does not ignore the role of government policy (whether wetlands or food security related), rather it emphasizes that local institutions should be allowed to play a more active role in mediating and developing policy in collabora-tion with other stakeholders rather than being at the end of the policy chain. This can similarly be applied to catchment-based institutions, as in the case of Wichi, where the ESS and livelihood dynamics between wet-lands and catchment are inextricably linked (see for example, Chapter 3). Second, this chapter has also argued that such local institutions can exist as hybrid structures that incorporate traditional cultural institutional val-ues while simultaneously working in cooperation with supportive external actors. Rather than regarding this as a failure of indigenous management, a more positive view is that it presents an opportunity for collaboration between different wetland stakeholders, which can create more diverse and resilient institutional arrangements. Third, and perhaps more impor-tantly, the research in western Ethiopia once again reaffirms the need for policy-makers to avoid a 'one size fits all' approach to local-level wetland management. There is little doubt that local institutional arrangements for wetland management in Ethiopia have been acutely defined by their socio-economic, cultural, environmental and political context; understanding these complexities should, therefore, be a matter of priority for wetland policy-makers.

Acknowledgements

The authors are grateful to the many farmers in Illubabor and Western Wellega who gave up their time to participate in the research on which this chapter is based. Thanks also go to the various wereda and zonal admin-istrative staff who helped facilitate both research and implementation activities. Research funding was provided by the British Academy, the UK Economic and Social Research Council, and the European Union. The Wichi Integrated Wetland and Watershed Natural Resources Management Project was funded by the Swedish International Development Co-operation Agency (SIDA) and the Sustainable Land Use Forum (Ethiopia).

References

Abebe, Y. D. (2003) 'Wetlands of Ethiopia: an introduction', in Y. D. Abebe and K. Geheb (eds) *Wetlands of Ethiopia: Proceedings of a Seminar on the Resources and Status of Ethiopia's Wetlands*, IUCN, Nairobi.

Adger, W. N. (2003) 'Social capital, collective action and adaptation to climate change', *Economic Geography*, vol. 79, no. 4, pp. 387–404.

Agrawal, A. (2001) 'Common property institutions and sustainable governance of resources', *World Development*, vol. 29, no. 10, pp. 1649–1672.

Asres, T. (1996) *Agro-Ecological Zones of South-West Ethiopia*, Addis Ababa University Press, Addis Ababa.

Bardhan, P. (1993) 'Symposium on management of local commons', *Journal of Economic Perspectives*, vol. 7, no. 4, pp. 87–92.

Blunt, P. and Warren, D. M. (1996) *Indigenous Organizations and Development*, ITDG Publishing, London.

Brokensha, D., Warren, D. and Werner, O. (1980) *Indigenous Knowledge Systems and Development*, University Press of America, Lanham.

Carney, D. (ed.) (1998) *Sustainable Rural Livelihoods: What Contribution Can We Make?*, DFID, London.

Central Statistics Agency (2010) *Statistical Abstract*, CSA, Addis Ababa.

Chambers, R. (1983) *Rural Development: Putting the Last First*, Longman, London.

Chambers. R., Pacey, A. and Thrupp, L. A. (1989) *Farmer First: Farmer Innovation and Agricultural Research*, ITDG Publishing, London.

Cox, M., Arnold, G. and Tomás, S. V. (2010) 'A review of design principles for community-based natural resource management', *Ecology and Society*, vol. 15, no. 4, art. 38.

Craig, G. and Mayo, M. (eds) (1995) *Community Empowerment: A Reader in Participation and Development*, Zed Books, London.

Crewett, W., Bogale, A. and Korf, B. (2008) *Land Tenure in Ethiopia: Continuity and Change, Shifting Rulers, and the Quest for State Control*, CAPRi Working Paper no. 91, IFPRI, Washington, DC.

Dasgupta, P. (2003) 'Social capital and economic performance: analytics', in E. Ostrom and T. K. Ahn (eds) *Foundations of Social Capital*, Edward Elgar, Cheltenham.

Dixon, A. B. (2003) *Indigenous Management of Wetlands: Experiences in Ethiopia*, Ashgate, Aldershot.

Dixon, A. B. (2005) 'Wetland sustainability and the evolution of indigenous knowledge in Ethiopia', *The Geographical Journal*, vol. 171, no. 4, pp. 306–323.

Dixon, A. B. (2008) 'The resilience of local wetland management institutions in Ethiopia', *Singapore Journal of Tropical Geography*, vol. 29, no. 3, pp. 341–357.

Ellis, F. (2000) *Rural Livelihoods and Diversity in Developing Countries*, Oxford University Press, Oxford.

Ethiopia Network on Food Security (2001) 'Assessment to food deficit areas of North Omo and bonei wetlands of Welega and Illubabor finds good production', http://reliefweb.int/sites/reliefweb.int/files/resources/369D9770928FFD 7085256A1800542553-usaid_eth_15mar.pdf, accessed 9th February 2012.

Fisher, R. J., Magginnis, S. and Jackson, W. J. (2005) *Poverty and Conservation: Landscapes, People and Power*, IUCN, Gland.

Folke, C., Carpenter, C., Elmqvist, L., *et al.* (2002) *Resilience and Sustainable Development: Building Adaptive Capacity in a World of Transformations*. Scientific Background Paper on Resilience for the process of The World Summit on Sustainable Development on behalf of The Environmental Advisory Council to the Swedish Government, Norsteds Tryckeri AB, Stockholm.

Gole, T. W. (2003) *Vegetation of the Yayu Forest in SW Ethiopia: Impacts of Human Use and Implications for Situ Conservation of Wild Coffea Arabica L. Populations*, Ecology and Development Series no. 10, Centre for Development Research, University of Bonn, Bonn.

Grootaert, C. (1998) 'Social capital: the missing link?' *Social Capital Initiative Working Paper*, no. 3, The World Bank, Washington, DC.

Hailu, A., Abbot, P. G. and Wood, A. P. (2000a) *Appropriate Techniques for Sustainable Wetland Management*, Unpublished Report, University of Huddersfield, Huddersfield.

Hailu, A., Dixon, A. B. and Wood, A. P. (2000b) *Nature, Extent and Trends in Wetland Drainage and Use in Illubabor Zone, Southwest Ethiopia*, Unpublished Report, University of Huddersfield, Huddersfield.

Hassen, M. (1990) *The Oromo of Ethiopia: A History 1570–1860*, Cambridge University Press, Cambridge.

Hillman, J. C. and Abebe, D. A. (1993) 'Wetlands of Ethiopia', in J. C. Hillman (ed.) *Ethiopia: Compendium of Wildlife Conservation Information NYZS*, The Wildlife Conservation Society International, New York.

Hinchcliffe, F., Thompson, J., Pretty, J., Guijt, I. and Shah, P. (1999) *Fertile Ground: The Impacts of Participatory Watershed Management*, ITDG Publishing, London.

Howes, M. (1997) 'NGOs and development of local institutions: a Ugandan case-study', *The Journal of Modern Africa Studies*, vol. 35, no. 1, pp. 17–35.

Howes, M. and Chambers, R. (1979) 'Indigenous technical knowledge: analysis, implications and issues', *IDS Bulletin*, vol. 10, no. 2, pp. 5–11.

Koku, J. E. and Gustafsson, J. E. (2001) 'Local institutions and natural resource management in the South Tongu District of Ghana: a case study', *Sustainable Development*, vol. 11, no. 1, pp. 17–35.

Leach, M. and Mearns, R. (eds) (1996) *The Lie of the Land: Challenging Received Wisdom on the African Environment*, James Currey, Oxford.

MA (Millennium Ecosystem Assessment) (2005) *Ecosystems and Human Well-Being: Wetlands and Water Synthesis*, World Resources Institute, Washington, DC.

Manig, W. (1999) 'Have societies developed indigenous institutions enabling sustainable resource utilization?' *Journal of Sustainable Agriculture*, vol. 14, no. 1, pp. 35–52.

Mazzucato, V. and Niemeijer, D. (2002) 'Population growth and environment in Africa: local informal institutions, the missing link', *Economic Geography*, vol. 78, no. 2, pp. 171–193.

McAslan, E. (2002) 'Social capital and development', in V. Desai and R. B. Potter (eds) *The Companion to Development Studies*, Arnold, London.

McCann, J. (1995) *People of the Plow: An Agricultural History of Ethiopia 1800–1990*, University of Wisconsin Press, Wisconsin.

Mearns, R. (1995) 'Institutions and natural resource management: access to and control over woodfuel in East Africa', in T. Binns (ed.) *People and Environment in Africa*, John Wiley, Chichester.

Ostrom, E. (1990) *Governing the Commons: The Evolution of Institutions for Collective Action*, Cambridge University Press, Cambridge.

Ostrom, E. (2000) 'Social capital: a fad or a fundamental concept?' in P. Dasgupta and I. Serageldin (eds) *Social Capital: A Multifaceted Perspective*, World Bank, Washington, DC.

Pelling, M. and High, C. (2005) 'Understanding adaptation: what can social capital offer assessments of adaptive capacity?' *Global Environmental Change*, vol. 15, no. 4, pp. 308–319.

Pretty, J. (1995) 'Participatory learning for sustainable agriculture', *World Development*, vol. 23, no. 8, pp. 1247–1263.

Pretty, J. and Ward, H. (2001) 'Social capital and the environment', *World Development*, vol. 29, no. 2, pp. 209–227.

Putnam, R. (1995) 'Bowling alone: America's declining social capital', *Journal of Democracy*, vol. 6, no. 1, pp. 65–78.

Quibria, M. G. (2003) 'The puzzle of social capital: a critical review', *Asian Development Review*, vol. 20, no. 2, pp. 19–39.

Quirin, C. and Dixon, A. B. (2012) 'Food security, politics and perceptions of wildlife damage in Western Ethiopia', *International Journal of Pest Management*, vol. 58, no. 2, pp. 101–114.

Rasmussen, L. N. and Meinzen-Dick, R. (1995) *Local Organisations for Natural Resource Management: Lessons from Theoretical and Empirical literature*, EPTD Discussion Paper no. 11, International Food Policy Research Institute, Washington, DC.

Reij, C. and Waters-Bayer, A. (eds) (2001) *Farmer Innovation in Africa: A Source of Inspiration for Agricultural Development*, Earthscan, London.

Ribot, J. C., Chhatre, A. and Lankina, T. V. (2008) Institutional Choice and Recognition in the Formation and Consolidation of Local Democracy, Representation, Equity and Environment Working Paper no. 35, World Resources Institute, Washington, DC.

Richards, M. (1997) 'Common property resource institutions and forest management in Latin America', *Development and Change*, vol. 28, no. 2, pp. 95–117.

Richards, P. (1985) *Indigenous Agricultural Revolution: Ecology and Food Production in West Africa*, Hutchinson, London.

Scoones, I. (1998) 'Sustainable rural livelihoods: a framework for analysis', *IDS Working Paper*, no. 72, Institute of Development Studies, Brighton.

Serra, A. (2001) *Legitimacy of Local Institutions in Natural Resource Management: The Case of Pindangaga, Mozambique*, Marena Research Project Working Paper no. 2, School of Africa and Asian Studies, The University of Sussex, Falmer.

Shivakumar, S. J. (2003) 'The place of indigenous institutions in constitutional order', *Constitutional Political Economy*, vol. 14, no. 1, pp. 3–21.

Swift, J. J. (1979) 'Notes on traditional knowledge, modern knowledge and rural development', *IDS Bulletin*, vol. 10, no. 2, pp. 41–43.

Tato, K. (1993) *Evaluation of the Environmental Components of Menschen für Menschen Projects in Ethiopia*, Unpublished Report, MFM, Addis Ababa.

Tiffen, M., Mortimore, M. and Gichuki, F. (1994) *More People, Less Erosion: Environmental Recovery in Kenya*, John Wiley, Chichester.

Uphoff, N. (1992) *Local Institutions and Participation for Sustainable Development*, Gatekeeper Series no. 31, IIED, London.

Wade, R. (1988) *Village Republics: Economic Conditions for Collective Action in South India*, ICS Press, Oakland.

Warren, D. M., Slikkerveer, L. J. and Brokensha, D. (1995) *The Cultural Dimension of Development: Indigenous Knowledge Systems*, ITDG Publishing, London.

Watson, E. (2003) 'Examining the potential of indigenous institutions for development: a perspective from Borana, Ethiopia', *Development and Change*, vol. 34, no. 2, pp. 287–309.

Woldu, Z. (2000) *Plant Biodiversity in the Wetlands of Illubabor Zone*, Unpublished Report, University of Huddersfield, Huddersfield.

Wood, A. P. (1996) 'Wetland drainage and management in south-west Ethiopia: some environmental experiences of an NGO', in A. Reenburg, H. S. Marcusen and I. Nielsen (eds) *Proceedings of the Sahel Workshop 1996*, University of Copenhagen, Copenhagen.

Wood, A. P. and Dixon, A. B. (eds) (2002) *Sustainable Wetland Management in Illubabor Zone: Research Report Summaries*, Wetlands and Natural Resources Research Group, University of Huddersfield, Huddersfield.

5 The emergence of a systemic view for the sustainable governance and use of wetlands in complex and transforming environments

Experiences from Craigieburn, South Africa

Sharon Pollard and Derick du Toit

Summary

This chapter explores the development of a systemic approach for sustainable wetland use against a history of research and development in the Manalana wetland of Craigieburn village where – as in many former Bantustans of the apartheid era – people use wetlands to sustain their livelihoods. However, degradation is widespread and socio-political factors weaken the potential for long-term sustainability. The focus of the work was to explore how the challenges of sustainability and livelihood security could be met within the context of political transformation and development. A central thesis is that wetland degradation and livelihood security need to be viewed as part of a wider socio-ecological and political system if realistic solutions are to be sought. These are both systemic issues requiring a systemic response. Understanding which drivers (including poverty, governance and social capital) influence system dynamics is vital to planning for change. Emerging from this work has been a deeper conceptualisation of the term '*wise use*'. In Craigieburn it was concluded that governance (locally tenable, reflexive) and social capital (such as trust, identity and the ability to self-organise) lie at the heart of wise use. Given the livelihood dependency on wetlands such as Craigieburn the impetus for substantive change (beyond improved individual practice) is evident but remains elusive, and reasons for this are discussed.

Introduction

The village of Craigieburn is similar to many of the eighty villages scattered throughout the Sand River Catchment in the north-eastern region of South Africa (Figure 5.1). Largely products of the apartheid regimes' policies of separate development, betterment and villagisation from the 1950s until democracy in 1994, the legacy of that era is still pervasive (Pollard *et al.*, 2003). High levels of unemployment, attendant poverty, poor education and alarming rates of HIV-AIDS render the livelihoods of most people extremely vulnerable and today people are largely reliant on wage remittances and government grants for cash. A striking feature of the catchment is the dense concentrations of people (150–350 persons/km^2) in so-called 'rural' areas juxtaposed with sparsely-settled, often affluent areas. This is starkly evident in Craigieburn where the population increased by 1,000 per cent in just nine years between 1965 and 1974 as a result of forced migration for the creation of Bantustans (or so-called 'self-governing states') for the majority of the black population (Pollard *et al.*, 2005). Under such conditions people turned to the land – and wetlands – to eke out a living, but a combination of low and erratic rainfall, poor soils and high numbers of people made any chance of developing a productive and sustainable, agricultural-based livelihood highly unlikely. Although the catchment is severely degraded (Pollard *et al.*, 1998), some natural resources, especially wetlands, offer an important contribution to peoples' livelihoods both directly and indirectly (Pollard *et al.*, 2008a). Wetlands in particular are valued for their relatively moist conditions – making them amenable to specific plants and crops such as taro (*Colocasia esculenta*) – and the promise of an extended grazing and growing season beyond that of rainfed uplands. However, wetlands are vulnerable systems in their own right in that the combination of high numbers of people, intensive use, highly erosive soils and their position in the landscape make them particularly susceptible to erosion and degradation.

The Manalana wetland used by the villagers of Craigieburn is a case in point (Figure 5.1). The wetland is used for a variety of activities including cultivation, livestock grazing and watering, reed harvesting and domestic water supply (Pollard *et al.*, 2005). However, the wetland is threatened by gulley erosion arising from a combination of factors including land use practices, both wetland and hillslope, exacerbated by the steep longitudinal wetland slope and sandy soils.

It is in this context that the work on wetlands and livelihoods in Craigieburn was undertaken, framed by a central question: *how could the challenges of sustainability and livelihood security be met, particularly within the context of political transformation and the imperative for development?* Is the 'balance' so often referred to in the sustainability literature and policy actually tenable? Whilst the focus was on Craigieburn wetlands, the issues at hand are of far wider applicability – for not only is Craigieburn but one village in the former Bantustans of South Africa – but also the issues of sustainable use and sustainability remain an 'elusive goal' (Walker and Salt, 2006). Thus, one of the objectives was to reflect on and learn from the work. This has wider implications, for like many developing countries, South Africa is committed to the constitutional principles of equity and sustainability as espoused in the National Water Act (RSA, 1998). Yet the vision of livelihood and wetland sustainability is a challenging one when seen in the wider context of huge social inequities and injustices of the past. Politically the incumbent government is challenged to deliver on the promises of jobs,

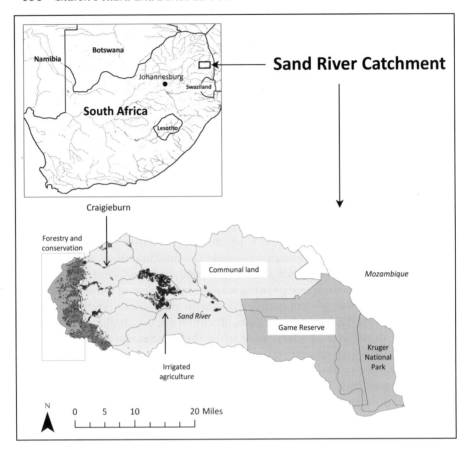

Figure 5.1 Map of the Sand River Catchment showing the location of Craigieburn,
      site of the Manalana wetland and the surrounding land use and land
      tenure arrangements

education, health and land reform, and given such pressing priorities it is
perhaps unsurprising that issues of sustainable wetland use receive scant
attention.

   This chapter explores the development of a systemic approach to wetland
sustainability against a history of research and development in Craigieburn.
Central to this is the argument that wetland degradation and livelihood
security need to be viewed as parts of a wider socio-ecological and political
system if realistic solutions are to be sought. We suggest that generic calls
for 'sustainable use', 'stakeholder participation' and 'wise use' in wetland
management are often based on a naïve conceptualisation of reality, which
ignores the wider – often nuanced and complex – socio-political, economic and environmental milieux. Equally, understanding which drivers

influence system dynamics is vital to planning for change. Emerging from this work a deeper conceptualisation of the term 'wise use' has arisen, an issue we return to later. This is important because despite Ramsar's attempt to recognise the nexus between wetlands and peoples' livelihoods (see Chapter 1), this has received insufficient attention to date, with some exceptions (see for example Pirot *et al.*, 2000; Millennium Ecosystem Assessment, 2003; Wood and van Halsema, 2008).

This chapter starts by providing a brief overview of the key conceptual frameworks that have been central to the work in Craigieburn. It then goes on to describe the development and deepening of an understanding of wetlands as complex systems and what this means for charting potential ways forward. This provides the context and motivation for the emergent concept and understanding of *wise use* which is introduced in the final section.

Conceptual grounding

Whilst a detailed description of the conceptual framework supporting this work is beyond the scope of the chapter, a number of key concepts merit introduction. Central are complexity theory and systems thinking which are briefly reviewed.

It is now increasingly recognised that variation and uncertainty are inherent properties of complex, linked, socio-ecological systems (SES) (Berkes and Folke, 1998) such as wetlands and their catchments. These ideas are not new, reflecting the cornerstones of systems theory (von Bertalanffy, 1968; Checkland, 1981; Forrester, 1992) which examined 'wholeness' and connectedness. In recognising wholeness or systems, there was also the recognition that their behaviour is often unpredictable because of the variable effects of cross-scale, multiple drivers (usually only a few dominant ones), thresholds (sudden sharp rather than linearly progressive change) and feedbacks (either dampening or exacerbating reactions). Another important property of complex systems is *emergence* where feedbacks generate surprising new properties not predictable from the original components making up the system. The concern for those interested in SES was that the reductionist approaches of conventional scientific method which had so influenced natural resource management (NRM) were based on averages (e.g. harvesting according to a maximum sustainable yield despite variations in the wider context). These are considered to be ill-equipped to deal with complexities found in reality (see for example Levin, 1999; Holling, 2001; Folke *et al.*, 2002; Gunderson and Holling, 2002; Walker *et al.*, 2004). In fact, many assert that despite the enormous investment in the science and management of natural resources, sustainability crises have arisen precisely because NRM approaches (informed by science) are not sensitive to spatial and temporal variability (Walker and Salt, 2006). Moreover, Heylighen *et al.* (2007) point out that from a philosophical point of view, once human agency and freewill is included, there is a basic contradiction with determinism. Indeed, classical science ignores issues of ethics and values (see for example Cilliers, 1998), and yet how systems are managed and used is ultimately determined by people. It is increasingly evident that variation and dynamism are properties that confer system resilience through the ability to respond, learn and adapt and hence should be fostered, not ignored (Cilliers, 1998; Holling, 2001; Ison, 2010).

This shift in thinking poses new challenges for governance and management, not least of which is the discomfort of many politicians and practitioners with the idea that they cannot set a goal and a series of blueprint steps that guarantee some prescribed outcomes. Nonetheless, proponents of complexity theory and systems approaches recognise that despite unpredictability, management actions do need to be taken (Johnson, 1999). Adaptive management, in which one learns whilst doing and adapts management processes and goals, is regarded by many as the natural corollary for dealing with uncertainty (see for example Biggs and Rogers, 2003). An important property of complex systems is the role of feedbacks (Holland, 1999; Heylighen *et al.*, 2007; Pollard and du Toit, 2011), which occur when causal chains close on themselves feeding back to the starting condition (such as poverty leading to an increased drop-out rate at school and increased unemployment feeding back to poverty). These are regarded as important with respect to the resilience of a system (either in a positive or negative sense) and may even lead to persistent re-inforcing feedback loops that prevent systems from responding to change and so reduce their resilience in the long term (Allison and Hobbs, 2004).

Initial steps towards an integrated approach to wetland assessment/management

It is argued that adopting a systems view has been critical in piecing together the 'puzzle of sustainable use' in Craigieburn. Nonetheless this view emerged as part of an iterative process of research, learning and reflection that evolved over two phases, summarised in Table 5.1. Each phase entailed different elements, each with specific methods, although there was significant overlap. The theme of wetland health and livelihood security provided coherence across the phases, and an evolving systems analysis provided the framework for integration.

Phase I in Craigieburn started in 2003 (Table 5.1) under the banner of 'wetland health and livelihood security' – then a relatively new framing of the problem of wetland degradation (Pollard *et al.*, 2005), since no other studies in southern Africa were explicitly exploring the links between livelihood security and wetland degradation. At that time, understanding degradation was influenced by a linear, one-way conceptualisation with wetland degradation seen as resulting from *within wetland practices* alone (Figure 5.2). The response was therefore to address land use practices *within the wetlands* themselves and, in some cases, implement physical

Table 5.1 A summary of the work phases, key elements, their purpose and methods for Craigieburn

Phases, key elements and purpose	Overview of methods and participants/partners

Phase I: Understanding context and forming partnerships
Phase I sought to understand the biophysical, ecological and livelihoods role of the communal wetlands of the upper Sand River Catchment (Craigieburn) particularly in terms of water security. It involved baseline research on geomorphology, hydrology and vegetation ecology and on livelihoods and well-being of Craigieburn residents.

Geomorphology	• Description of basin shape and longitudinal profile (influences hydraulic properties of small catchments, particularly velocity of surface flow, and hence ability of water to erode). • Delineated wetland boundary and extent. • Analysis of soil chemistry from upslope, midslope and streambed to understand erodibility.
Vegetation	• Species composition determined at points along transects where soil measurements and elevation were measured. • Vegetation classification achieved using TWINSPAN. • Relationships between vegetation distribution and environmental characteristics described (edaphic indicators of wetness).
Hydrology	ACRU(Agricultural Catchments Research Unit) – modelling system (Schulze, 1989) using a three-tiered approach: 1 A plot-scale water balance study for different land uses on different soils. 2 Micro-catchment scale modelling (Craigieburn wetland and upstream area), to investigate changes in land use and wetland degradation. 3 Sand River Catchment-scale modelling exercise to investigate the potential changes arising from wetland degradation in the upper reaches of the catchment.
Livelihoods and well-being	Participatory well-being and livelihoods assessment using a livelihoods framework and locally established categories of well-being. Involved: 1) workshop with institutional structures; 2) two-day workshop with wetland users; and 3) semi-structured interviews with each household.
Preliminary assessment of governance	Participatory workshop of leadership and residents to understand existing institutional arrangements, and roles and responsibilities with regard to wetlands.
Systems analysis	Systems approaches and SES framework used to analyse key drivers and outcomes. Results shared and discussed with residents and other potential partners such as the Department of Agriculture.

Continued

Table 5.1 A summary of the work phases, key elements, their purpose and methods
for Craigieburn, *continued*

Phases, key elements and purpose	Overview of methods and participants/partners

Phase II: Being responsive
Phase II included various initiatives to address issues that emerged from Phase I
in an integrated way including in-depth hydrological analysis, a Farmer Support
Programme (FSP), support for improved governance and indicators project.

FSP	Worked with 120 farmers over three years, using action-research (see text) to support good practice for farming through:

1 collaborative assessment of land use practices;
2 collaboratively defined action projects;
3 farmer self-assessment (see indicators below) of the impact of their practices on production and wetland health;
4 improved capacity of agricultural extension officers.

Improved hydrological analysis	Research to determine: a) the wetland's hydrodynamic behaviour and the extent to which this had degraded as a result of erosion; and b) if technical rehabilitation has ameliorated any degradation in the wetland's hydrological condition. Hydrological monitoring of the catchment was undertaken with a network of groundwater piezometers and soil moisture tensiometers installed at various depths.
Integrated analysis of impacts of land use practices on wetland integrity	In order to explore the relative importance of different land use on wetland integrity, a dynamic model was developed using STELLA (tool for constructing understanding about dynamic systems, http://www.iseesystems.com). This work used existing data from the biophysical research (above) to explore the individual and cumulative impacts of changing certain land use practices such as ridge furrow (see text). It involved participation of all research team members.
Physical rehabilitation	The rehabilitation of two major headcuts in order to raise the water table and halt erosion was undertaken by Working for Wetlands (see text). There was little involvement of the local community – a key reason for the later development of the Wise Use Programme. The upland soil and water conservation initiative worked with farmers to demonstrate different practices. Long-term intervention constrained due to limited funding.

Continued

Table 5.1 A summary of the work phases, key elements, their purpose and methods for Craigieburn, *continued*

Phases, key elements and purpose	Overview of methods and participants/partners
Valuing ESS	ESS associated with wetlands were valued under two scenarios (with and without physical rehabilitation; see text for details). Involved: a) establishing the links between rehabilitation and the ecosystem secured (benefits); b) establishing the links between these benefits and peoples' livelihoods; and c) valuing these using a cost–benefit analysis.
Sustainability indicators in communal wetlands	Focused on the practical application by practitioners including the wetland users, and custodians of wetlands. Indicators developed collaboratively with users involving an iterative process of: a) describing the context; b) agreeing on key strengths and problems; c) agreeing on principles; d) developing a collective vision and objectives; and e) developing and testing associated indicators.
Governance	Research undertaken to explore the realities, needs, constraints and opportunities with regard to strengthening governance for sustainable wetland use. Participatory research (focus group meetings, key informant interviews) to develop a collective understanding of the past and present land management arrangements and future needs. Tested potential governance models for Craigieburn based on collaboratively developed vision, principles and an elaboration of rights, responsibilities, authority and the distribution of risks and benefits.
Systems analysis of key drivers and outcomes as above	As above but deepening the analysis by drawing on research and development through Phase II.

remediation measures. In the case of Craigieburn it was recognised that little was known about either the social or biophysical context, both essential for the development of a long-term management plan. Thus, the initial focus was on understanding the role of wetlands in peoples' livelihoods and the biophysical setting of the wetlands (namely geomorphology, hydrology and vegetation ecology). The results are summarised in Table 5.2 and specific issues elaborated below.

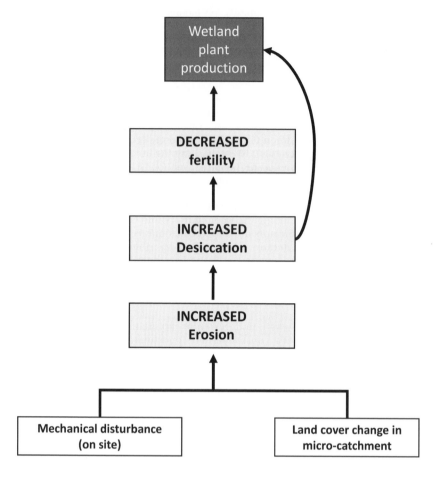

Figure 5.2 Simplified schematic of the links between land use and biophysical factors of erosion, desiccation and productivity

Note: This initial view was later expanded to a systemic representation, see Figure 5.4.

Table 5.2 Summary of the main biophysical and socio-economic characteristics of the study area; data specific to Craigieburn were collected during Phase 1

Characteristic	
Catchment context – Sand River Catchment	• Juxtaposition of wealthy and poor in the catchment as a whole. Population is largely rural; most of poor areas are former apartheid Bantustans (into which the majority of the black population was moved under the apartheid regime) characterised by major socio-economic problems.
Topography and rainfall	• Varies from mountainous areas of high rainfall (1,500 mm) to the low eastern plains with rainfall of less than 450 mm. • Craigieburn rainfall – 750 mm p.a. • Rainfall is seasonal, occurring mainly during summer (October–March).
Water resources	• Highly stressed; water requirements exceed water availability at the catchment level.
Land water use	• Land under communal tenure. • Subsistence, dryland agriculture is prevalent.
Catchment area	• Sand River Catchment – 2,000 km². • Craigieburn micro-catchment – 261 ha (2.6 km²).
Wetland description	• Mainly valley-head wetland, fed by hillslope seepage.
Wetland area	• 6% or 1,200 ha (12 km²) of Sand River Catchment. • Manalana wetland – 7.9 ha. • Decrease in extent since 1960 – 8% to 3% of Craigieburn micro-catchment area (20.88 ha to 7.83 ha today; Riddell *et al.*, 2012).
Wetland uses	• Primary – small-scale agriculture, principally taro (*Colocasia esculenta*). • Use of natural products, e.g. reeds (*Schoenoplectus corymbosus, Cyperus latifolius*), clay and springs. • Grazing and watering for livestock. • Supplementary domestic water supply. • Used almost exclusively for domestic purposes.
Wetland use	• Area of wetland plots ~ 6.4 ha (most of area suitable for cultivation is used). • Mean plot size of 0.36 ha, median of 0.23 ha.
Demographics of users and use	• Total no. of households = 182 (Craigieburn population ~ 1,300). • Number of female-headed households (single, adult-headed) – 53%. • Number of households using wetlands – 104 (60%). • Women of between 35 and 70 years of age, mainly from single-headed households.

Continued

Table 5.2 Summary of the main biophysical and socio-economic characteristics of the study area, *continued*

Characteristic	
Livelihoods of wetland users	• Generally livelihoods are very vulnerable with 25% of wetland users having minimal sources of food other than that from the wetland. • 61% have limited income (pensions, grants and occasional work, many dependents). • Only 14% are well-off (a family member with full-time job).
Livelihood contributions from wetlands	Estimated to contribute 40% of the food grown and offer an important safety net, particularly for the poor (Pollard *et al.*, 2008a).
Governance	General weakening in local-level governance with some 60% of households accessing land without any permission or negotiation, citing hunger as the key driver (see text).

Sources: Pollard *et al.*, 1998 and Pollard *et al.*, 2005

The wetlands are used principally by women, many of whom are household-heads and poor (Table 5.2). Based on increasing desiccation and declining productivity, users regarded the wetlands as badly degraded but felt that there were no alternatives for acquiring new land or addressing the degradation. Indeed, when engaged in participatory planning processes, few options for action and change were raised and a general sense of collective disempowerment was pervasive, with people saying 'there is nothing to do'. This suggested that the wetland-dependant households may be increasingly vulnerable, particularly in terms of food security, due to diminishing wetland health, a lack of individual and collective action and dwindling institutional controls (see later).

An examination of land use practices suggested that these were unsustainable. Although some positive practices existed, such as manual tillage and non-usage of agro-chemicals, a range of deleterious practices – mainly the absence of soil and water conservation practices – was observed (Table 5.3). The most prevalent degrading practices for wetland integrity were the ridge furrow systems including removal of vegetation and the poor orientation of raised beds which, when combined with the steep gradient of many drainage furrows, led to over drainage. Also of concern was the extent of cultivation taking place, with very few areas – especially sensitive zones like those near the toe of the wetland – being left intact. Farmers explained that when they were moved into the area they had little experience of wetlands but adapted farming practices from what they knew. They were drawn to the wetlands for reasons already explained – principally wetness – but then found the wetlands to be 'too wet' and hence developed systems to drain them.

Table 5.3 Summary of negative agricultural practices within the wetlands at Craigieburn

Features of the agricultural system	General impacts of the practice on: • Short-term productivity (italicised) • State of health of the wetland	Prevalence
Poor **orientation** of raised beds (i.e. parallel with water flow) and **gradient** of furrows too steep (i.e. >1%)	Excessive drainage (especially in dry periods) leads to: • *Potential water stress of crops.* • Desiccation of wetland and increased surface water velocity (especially stormflows), increased erosion and diminished sediment trapping.	Very high
Failure to **block furrows** and re-wet plot when fallow	Potential benefits of resting are not fully realised because conditions most favourable for the recovery of SOM (soil organic matter) are not created.	Very high
Lack of productive use of **furrow** areas	*Productivity of the overall plot is lower.*	High
Frequent and extensive **disturbance** through tillage	Exposes the wetland to greater risk from erosion and more rapid depletion of SOM.	High
Disturbance immediately **adjacent to a stream**	• *High risk of loss of the crop in a storm event.* • Increase in the risk of accelerated stream bank/gulley erosion.	Moderate
Low plant **density** especially during dry periods	• *Weed control more difficult; lower yield.* • Low vegetation cover provides less effective protection against erosion.	Moderate
Extensive cultivation – little area under natural vegetation	• Capacity to decrease surface water velocity and control erosion is compromised. • Value for biodiversity is greatly diminished.	High
Poor **weed** control	*Potentially considerable competition with crops, diminishing yield.*	High
Limited **resting** of the land	Limited opportunity for the recovery of SOM and trapping of fine sediment.	High
Burning of plants	Reduced amount available to contribute to SOM.	High
Crops with **high water demand** (e.g. sugar cane)	Desiccation of the wetland.	Moderate
Poor **manure** protection and application practices	• *Reduced amount of nitrogen available for crop growth.* • Reduced productivity.	High

Source: adapted from Pollard *et al.*, 2005

Farmers raised three priority concerns with the project team: desiccation of the wetlands, poor fertility and a decrease in productivity over the years. These were critical for the research in that they formed the basis of a conceptual, systemic model of the interaction of biophysical factors leading to degradation (see Figure 5.2). The biophysical team (a hydrologist, geomorphologist, vegetation ecologist and wetland specialist) examined these factors not only to understand each issue but also to explore causal relationships between them. The aim was to determine the geomorphological setting of Manalana wetland, without which it would be difficult to predict wetland function and hence sustainable management (McCarthy and Hancox, 2000; Ellery, 2005; Box 5.1).

Box 5.1 Understanding the desiccation of the Craigieburn landscape: the links between geomorphology, hydrology and erosion

The desiccation of a wetland may result from several different on-site (wetland) or micro-catchment factors. Two key interacting, on-site factors are the inherent soil properties (sandy dispersive soils with areas of high sodium concentrations, making them prone to erosion) and the presence of erosional gullies or artificial drainage channels (Ellery, 2005) (see Table 5.3). Moreover, the wetlands occur on steep slopes where the water table is held in place by a plug of fine clays at the toe of the wetland. Their position makes them vulnerable to headward erosion and when these clays are lost through gulley erosion, hydraulic conductivity and groundwater flow increases within the soil so that there is a gradual drawdown of the water table. This exerts an overall desiccating effect on the micro-catchment creating conditions which are unfavourable for the production of organic carbon, and fertility declines. Indeed, the vegetation analysis (decrease in wetland-dependent terrestrial species) suggested that there had been widespread desiccation of the landscape, and a 50 per cent reduction in wetland areas over about 20 years (Ellery, 2005).

The poor state of the micro-catchment (inadequate vegetation, extensive footpaths) acts to increase water velocities and concentrate runoff which, together with dispersive soils, contributes to erosion. Impacts include: a) desiccation of the wetland; and b) increased losses of organic matter, nutrients and sediment from the wetland through erosion.

The hydrological assessment highlighted the dearth of wetland hydrology modelling systems applicable at small scales. Nonetheless, the role of wetlands, such as Craigieburn, appears to be significant at the micro-catchment scale where they attenuate wet-season floods (Riddell *et al.*, 2012). However, the role of these wetlands in baseflow augmentation remained inconclusive.

Understanding the issue of desiccation (Box 5.1) proved to be fundamental for understanding degradation systemically not only because it highlighted that the source of erosion could be distant from the impacts (*upstream–downstream linkages*) through the upstream propagation of headward erosion in the stream channel, but also because it suggested that erosion within the wetland was having a *landscape scale* effect and vice-versa, through the influences of the surrounding micro-catchment on wetlands (*upland–wetland linkages*). It also explicitly demonstrated linkages between desiccation and soil fertility (mainly through the loss of organic matter) and hence loss of productivity.

The initiation of a systemic view represented a breakthrough in many ways. Not only did it facilitate engagement between different disciplines in the team who elaborated the links between the problems that farmers raised (Figure 5.2), but it also provided the basis for a more holistic understanding and collaborative planning with the community (see Figure 5.4).

Although like studies elsewhere on dambos (McCartney, 2001; Von der Heyden and New, 2003), the recent hydrological analysis (Riddell *et al.*, 2012) suggested that such systems can – given correct agronomic practices and the protection of certain vulnerable areas (such as the clays) – be sustainably utilised with little impact on river hydrology, shifting management and practice proved more challenging in reality. This phase of the work noted that whilst the Craigieburn wetlands clearly contribute to the livelihoods of the rural poor, theoretically providing a strong impetus to improve practices, the systemic analysis suggested that it is the very cycle of poverty coupled with weak governance, varying levels of awareness regarding wetlands, poor wetland practices and the lack of alternatives that confounds sustainable wetland use.

Deepening of a systems view

Emerging from this work were the cornerstones of a more systemic approach that involved greater collaboration with the community (see Table 5.1). These included: a) the rehabilitation through halting current gulley erosion in the Manalana wetlands and the securing ecosystem services (ESS); b) farmer support for sustainable wetland use practices through action-research and learning; and c) wider-scale issues of governance for NRM. Each of these is briefly discussed with an emphasis on the key features and major challenges that were so instrumental in the emerging systemic view.

Rehabilitation by halting current gulley erosion in the Manalana wetlands and securing ESS

The research revealed that without rehabilitation, the erosion headcuts would advance upstream through the remaining intact areas of the wetland

(Kotze *et al.*, 2008), impacting on an array of ESS enjoyed by local residents (water, harvestable products, crops, grazing). Therefore, Working for Wetlands, a national programme aimed at the physical rehabilitation of wetlands (http://wetlands.sanbi.org) rehabilitated two major gulley-erosion headcuts so as to raise the water table. The work has restored the wetlands artesian hydrodynamics to that typical of conditions upstream of a clay-plug (Riddell *et al.*, 2012). Moreover an assessment of the ESS estimated that the rehabilitation had secured a four-fold net benefit of R149,256 (US$21,730) per year in comparison to R38,196 (US$5,300) under degraded conditions (Pollard *et al.*, 2008a).

Farmer support for sustainable wetland use practices through action-research and learning

The FSP was initiated to support sustainable wetland practices through an action-research orientation to understanding with farmers: why they do what they do; how they understand degradation and how they might respond appropriately to change and unpredictability that are intrinsic to the system. Through projects of experimentation and learning – framed by a collective, farmer vision – the FSP activities moved through stages from conceptual grounding to a more critical engagement with practical skills and action over three years (du Toit *et al.*, 2008). So as to encourage collaborative learning and action, some 104 farmers worked in clusters according to the location of their fields. Most farmers had clear ideas on sustainable practices but these were often piecemeal and varied between individuals. Thus one of the key challenges was designing an approach that could communicate an *integrated view* of the concept of sustainability and that could be shared by all farmers whilst addressing varying literacy levels. For example, the 'Hand of Change' (where every farmer has a hand, Figure 5.3) was a simple tool designed to organise complex principles and information associated with the suite of soil and water conservation practices that need to be considered within wetlands.

Special attention was paid to reflexivity in an ongoing, open-ended process of experiential learning, dialogue and reflection. A major issue that emerged for farmers was recognising their own lack of communication, self-organisation and trust (amongst themselves and with government officials). These issues highlight the cultural and social nature of farmer support that is needed, and concur with much of the literature that examines the links between sustainable NRM, strong social capital and community-based local institutions (see Chapters 1 and 4). Farmers noted the benefits of collaboration and continue to meet today, although less formally.

Three key issues emerged from the FSP initiative. First, despite the best intentions by farmers to address the negative land use practices (Table 5.3), the realities were different. Some improved practices that were relatively easy to adopt, such as mulching and protection of wetland plants, were taken up,

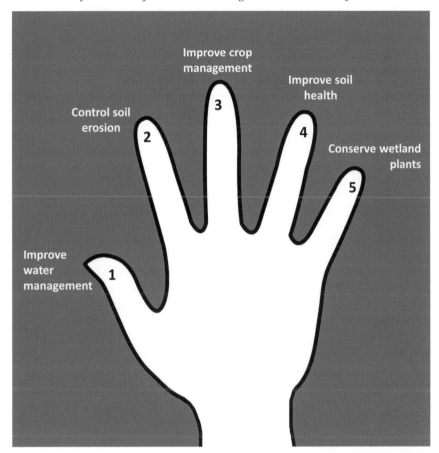

Figure 5.3 The 'hand of change' as an example of heuristics designed to help farmers think through key but complex soil and water conservation priorities for the sustainable use of wetlands. Each focus area (finger) elaborates the key problems noted in Phase I

whilst others (such as changing bed orientation), were regarded as too oner-ous or risky to attempt. Clearly, undertaking the full range of recommended changes to land use practices would be unlikely, especially for the elderly. Second, later research that examined the relative impact of changing prac-tices on wetland health suggested that farmers could have significant impacts on wetland integrity – and hence production – fairly simply through the application of manure, attention to mulching and fallowing of land (Pollard *et al.*, 2008c). Importantly, this also suggested that wetland health could be secured in different and multiple ways, thereby questioning the notion of 'best practice' as a singular approach to achieving sustainability (Pollard *et al.*, 2009). Third, farmers also confronted problems that had their origins in the micro-catchment rather than in the wetland itself (e.g. livestock damaging crops, fire, gully erosion and sediment inputs from the surrounding villages). Not only did this require looking 'beyond their fence' but also illustrated that wetlands do not exist in a vacuum but are part of broader, mutually influencing systems (see Figure 5.4), and that governance – both statutory and 'customary' – lies at the heart of *how* wetlands – and the surrounding micro-catchment – are used (Pollard *et al.*, 2008a).

Governance at the heart of NRM in Craigieburn

With the need for improved governance (referring to rights, responsibilities and authority over natural resources) at multiple levels, a fairly ambitious initiative to *facilitate wise and effective governance of natural resources in Craigieburn* was developed subsequently. Over three years the action-research approach allowed for collaborative development of various conceptual and planning tools (Cousins *et al.*, 2009). These were used to examine the local governance systems as well as the evolving institutional environment within both the *de jure* and *de facto* customary and statutory context – a situation known as legal pluralism (Lewanika, 2002; Meinzen-Dick and Pradhan, 2002).

A number of issues were striking. Two constraints, including the weakening governance over natural resources that had developed since 1994 in communal areas[1] (Pollard *et al.*, 2005) combined with the uncertain and contested nature of land reform (Cousins, 2007), were seen as likely to reverberate in local efforts to manage natural resources. Further, the NRM system is fundamentally shaped by plural tenure systems so that rights and authority derive both from custom and from statutory laws. However, whilst customary rights are well understood locally, statutory rights are not (Cousins *et al.*, 2009).

It became clear that major local and systemic constraints to effective governance exist – not least of which is that natural resources are not prioritised and hence residents are apathetic about committing time and resources. Wetlands – as part of the commons – have weaker tenure arrangements than other land uses and the sense of autonomy around wetland use poses particular challenges for collective, community-based governance (Cousins *et al.*, 2009). Added to this are issues of poverty, power and contested discourses regarding responsibility and authority, and a lack of knowledge of the new statutory systems by locals (Pollard and Cousins, 2008). People still turn to the known system, albeit largely ineffectual, for direction and action. Overall the governance system appears to be weakening as a result of both internal and external drivers (including lack of clarity on policy reforms and contested authorities and usufruct interests).

Although a detailed explanation is beyond the scope of this chapter, support for improved governance was difficult, despite some significant progress, and the outcomes were somewhat discouraging (Pollard and Cousins, 2008). It required working collaboratively across different interests and authorities in an unclear policy environment, and extending the boundary for governance beyond the wetlands – both of which proved extremely challenging in practice. The fact that authorities at all levels failed to provide recourse compounded the sense of disempowerment, whilst for transgressors it provided a milieu of impunity. It was concluded that the dynamic, ambitious and under-capacitated policy context as well as the weakening and contested nature of governance offer little support for local efforts to govern natural resources (Cousins *et al.*, 2009). Without clear rights and responsibilities, the ability to self-organise and gain access to recourse, change seems highly unlikely in the near future. On the positive side, farmers have taken up issues of concern on a number of occasions suggesting that collective action is possible.

Re-conceptualising wetland degradation as a systemic issue

The development of a more systemic view of wetlands and livelihood security underscored a number of key issues as shown in Figure 5.4 and embraced many of the issues raised by wetland users themselves. First, the role of governance (both customary and statutory) is emphasised as a key driver in natural resource use. The best collective practices in the world will likely be subject to transgressions at one time or another, and if there is no effective system (governance) for dealing with this – either through regulation or coercion – sustainable wetland use is unlikely (Cousins *et al.*, 2009; see also Pollard and du Toit, 2011). Second, key feedbacks – such as that between livelihood security and land use – are critical for understanding the dynamics of sustainable use. Notably, under conditions of poverty, peoples' priorities are less likely to be concerned with long-term sustainability as opposed to putting food on the table (Agarwal, 2006; Shackleton *et al.*, 2008) but as livelihood security improves so might the possibilities of sustainable practices (Pollard *et al.*, 2008b). Other important, wide-scale drivers that are increasingly influencing the system dynamics are market forces, consumer values and social capital. Usufruct issues are relatively new but highly volatile as illustrated by a large brick factory next to the wetlands that, without due remediation measures, has led to major hill slope erosion, silting up of wetlands and community tensions through the promise of financial benefits and jobs that have not materialised. Here external entrepreneurs have taken advantage of the weakened governance systems (see Cousins *et al.*, 2009) to establish business interests that are not compliant with national legal requirements. As Figure 5.4 suggests, ultimately the most pernicious and sustained risks will be borne most directly by the community because of their direct dependence on wetlands.

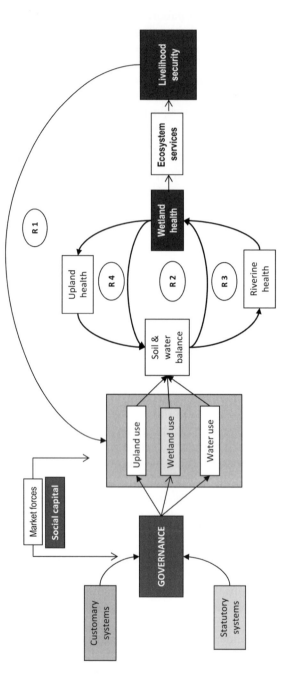

Figure 5.4 Systemic representation of wetland health and livelihood security (as a causal loop diagram) based on findings from Craigieburn

Note: the shaded box is where the original emphasis was placed but later expanded to include multiple factors. This shows the links between the governance, land and water use, wetland health and livelihood security, as well as key drivers and feedbacks, in this case as reinforcing loops (R). The model highlights the links between ecosystem integrity (wetland health) and livelihood security. Note that although livelihood security was the focus of rehabilitation, additional ESS are derived from healthy wetlands including biodiversity.

From this emerged a working definition of 'wise use' that is now being used by the national Working for Wetlands Wise Use Programme (Box 5.2). This builds on the current definition of Wise Use by Ramsar (2007) but is more explicit about issues of scale, the multiplicity of drivers and feedbacks, and the acceptance of uncertainty. Although the inclusion of the term *ecosystem approach*[2] in the Ramsar definition is suggestive of a wider, holistic approach, it falls short of recognising the interplay of wetland degradation and livelihood dependencies as systemic and hence complex, dynamic and often unpredictable. By implication this requires different approaches that are attentive to process, learning and adaptation. Moreover language can present a significant barrier or enabler of ideas (Stibbe, 2004) and, although not necessarily true, the term *ecosystem approaches* may well intimate that such approaches focus on conservation interests alone.

Box 5.2 Working definition of the wise use of wetlands as conceptualised by the Working for Wetlands Wise Use Programme

Source: Pollard and Cousins, 2010

Wise use is the sustainable use of wetlands through good stewardship and integrated land and water use practices that promote healthy wetlands so as to continue to provide services, products and benefits enjoyed by users and sustaining human livelihoods (including those of future generations), underpinned by biological diversity.

The wise use of wetlands recognises that wetlands are part of a broader socio-political and ecological milieu that is inherently dynamic and complex in nature. This position recognises that sustainable wetland use is part of a mutually influencing system of drivers, interrelationships and feedbacks requiring approaches to use and rehabilitation that are people-centered, flexible, adaptive and locally appropriate.

Conclusions: re-examining the meaning of sustainable use through a systemic lens

This chapter has argued that conventional perspectives on the links between wetland degradation and wetland practices reflect a narrow conceptualisation of 'the problem' that can result in equally narrow responses (and 'truth' claims) which, although tenable in their own right, may ultimately fail to secure livelihood benefits through sustainable use. In contrast, approaches that explicitly take account of the wider environmental, economic and socio-political milieux are more likely to be resilient and adaptive to change – an important attribute of complex systems. This does not imply that all wetland situations are complex (see Snowden, 2003), nor that every piece of wetland work must be systemic in nature since. As we have demonstrated, discipline-specific endeavours are important. However, it is incumbent upon researchers and practitioners to move from the naïve claims (such as 'awareness raising' as the silver bullet for poor land use practices) to a framework that facilitates thinking beyond the wetland in the search for 'sustainable solutions'. Moreover, there is no blueprint; each context has its own dynamic and nuanced socio-economic, political and environmental profile that offers both opportunities and constraints. As recognised by Wood and van Halsema (2008) in the Food and Agriculture Organization of the United Nations' (FAO) Guidelines for Agriculture-Wetland Interactions, the application of assessment tools such as the Drivers-Pressures-State Changes-Impacts-Responses (DPSIR) framework (see Chapter 1) offers one potential way forward for identifying such nuances, and locally suited management interventions. Another important framework is that being used by the Kruger National Park for identifying drivers, thresholds of potential concern and adaptive management approaches against a consensual vision (see Biggs and Rogers, 2003; Pollard and du Toit, 2007). Adopting a learning approach to management, such as that embodied in adaptive management, is seen as appropriate in an uncertain world (see for example Kingsford *et al.*, 2011).

As illustrated in the research in Craigieburn, such a view acknowledges a number of key factors. At the heart of these is the thesis that wetlands are complex in their own right and part of a wider, linked, complex and dynamic socio-economic, political and ecological system. Thus wetland degradation and livelihood vulnerability are systemic issues requiring a systemic response. Significant is the recognition that governance is central to determining *how* wetlands – and the surrounding micro-catchment – are used (Pollard *et al.*, 2008a). Given the livelihood dependency on the wetlands currently in Craigieburn, the impetus to improve practices and hence production exists and yet substantive change remains elusive. Farmers feel that even if they change their own practices, bringing others 'into line' within a common code (or set of rules) would be difficult given the lack of recourse and the contested nature of governance. In the current climate of change, where authorities and responsibilities remain unclear, the state seems unable to give tenable support in this regard. This sentiment of farmers is suggestive of a general feeling of disempowerment to act – either individually or collectively – outside of one's own field and points to the importance of social capital. Indeed work in Craigieburn suggests that governance together with strong social capital lie at the heart of sustainable use since, as Heylighen *et al.* (2007) point out, once human agency and freewill are considered, deterministic assumptions based on simplistic cause and effect are challenged. Developing a sense of custodianship over one's own natural resources depends on issues of trust, identity (as a custodian or simply a user) and the ability to organise and self-regulate (amongst a number of factors (Ison *et al.*, 2007; Pollard and du Toit, 2011). Thus, although one may work with farmers on an individual and collective basis, this wider context informs what can be done by whom and how. Although much remains to be addressed in Craigieburn in this regard, emergence, and hence unexpected outcomes or surprise, is a key property of complex systems (see earlier). For example, history teaches us that collaboration can and does emerge out of a historically fractious situation, especially in cases where resources are under increasing pressure (Ostrom, 1990; Agarwal and Angelsen, 2009), such as in Craigieburn. Importantly, because certain factors cannot be 'known' *a priori*, equipping people with a strongly reflexive and flexible process of management and action that is responsive and adaptable to change is needed if we are to give meaning to 'wise' and 'sustainable' use of wetlands in the future.

Notes

1 Prior to 1994 controls were effected through chiefs that were imposed by the white government, who dealt with transgressors through a system of 'tribal police'. After 1994 such individuals considered to be lackeys of the apartheid state were contested but this should not be conflated with a contestation of 'traditional systems' *per se.*
2 Ramsar (2007) updated their definition to include: 'the maintenance of their ecological character within the context of sustainable development, and achieved through the implementation of *ecosystem approaches*' (emphasis added).

References

Agarwal, A. and Angelsen, A. (2009) 'Using community forest management to achieve REDD+ goals 201', in A. Angelsen, M. Brockhaus, M. Kanninen, E. Sills, W. D. Sunderlin and S. Wertz-Kanounnikoff (eds) *Realising REDD+: National strategy and policy options*, CIFOR, Bogor, Indonesia.

Agarwal, R. M. (2006) 'Globalization, local ecosystems, and the rural poor', *World Development*, vol. 24, no. 12, pp. 1405–1418.

Allison, H. E. and Hobbs, R. J. (2004) 'Resilience, adaptive capacity and the 'lock-in trap' of the Western Australian agricultural region', *Ecology and Society*, vol. 9, no. 1, art. 3.

Berkes, F. and Folke, C. (1998) *Linking Social and Ecological Systems: Management Practices and Social Mechanism for Building Resilience*, Cambridge University Press, Cambridge.

Biggs, H. and Rogers, K. H. (2003) 'An adaptive system to link science, monitoring and management in practice', in J. T. D. Toit, K. H. Rogers and H. C. Biggs (eds) *The Kruger Experience: Ecology and Management of Savanna Heterogeneity*, Island Press, Washington, DC.

Checkland, P. (1981) 'Systems thinking, systems practice: Assessment tool for sustainable agroecosystem management', *Ecological Applications*, vol. 11, pp. 1573–1585.

Cilliers, P. (1998) *Complexity and Postmodernism: Understanding Complex Systems*, Routledge, London and New York.

Cousins, B. (2007) 'More than socially embedded: The distinctive character of "communal tenure" regimes in South Africa and its implications for land policy', *Journal of Agrarian Change*, vol. 7, no. 3, pp. 281–315.

Cousins, T., Pollard, S. R., Toit, D. D., Maboyi, J. and Wolf, J. D. (2009) 'Developing community based governance of wetlands in Craigieburn', *Technical Report to the IDRC*, June 2007–2009.

du Toit, D., Chuma, E., Makhabela, B. B. and Pollard, S. (2008) 'Building capacity for changed practices: An action-research project for wise use of wetlands as a contribution to an integrated rehabilitation and management plan for wetlands in the communal areas of the Sand River Catchment', *Final Report to World Wildlife Fund SA*, Green Trust 1368.

Ellery, W. (2005) 'Geomorphology of Manalana wetlands, Craigieburn. Linking water and livelihoods: The development of an integrated wetland rehabilitation plan in the communal areas of the Sand River Catchment as a test case', *The Save the Sand Programme*, AWARD, Prepared for WARFSA/Working for Wetlands.

Folke, C., Carpenter, S., Elmqvist, T., Gunderson, L., Holling, C. S. and Walker, B. (2002) 'Resilience and sustainable development: Building adaptive capacity in a world of transformations', *Ambio*, vol. 31, no. 5, pp. 437–440.

Forrester, J. W. (1992) 'System dynamics and learner-centred-learning in kindergarten through 12th grade education', Massachusetts Institute of Technology, www.clexchange.org/ftp/documents/sdintro/D-4337.pdf (accessed 17 October 2012).

Gunderson, L. H. and Holling, C. S. (eds) (2002) *Panarchy: Understanding Transformations in Human and Natural Systems*, Island Press, Washington, DC.

Heylighen, F., Cilliers, P. and Gershenson, C. (2007) 'Complexity, Science and Society', in J. Bogg and R. Geyer (eds) *Philosophy and Complexity*, Radcliffe Publishing, Oxford.

Holland, J. H. (1999) *Emergence: From Chaos to Order*, Perseus Books, Reading.

Holling, C. S. (2001) 'Understanding the complexity of economic, ecological and social systems', *Ecosystems*, vol. 4, no. 5, pp. 390–405.

Ison, R. (2010) *Systems Practice: How to Act in a Climate-change World*, Springer, London and Open University, Milton Keynes.

Ison, R. L., Röling, N. and Watson, D. (2007) 'Challenges to science and society in the sustainable management and use of water: Investigating the role of social learning', *Environmental Science and Policy*, vol. 10, no. 6, pp. 499–511.

Johnson, B. J. (1999) 'The role of adaptive management as an operational approach for resource management agencies', *Conservation Ecology*, vol. 3, no. 2, art. 8.

Kingsford, R., Biggs, H. and Pollard, S. (2011) 'Strategic adaptive management in freshwater protected areas and their rivers', *Biological Conservation – Special Issue Article: Adaptive Management for Biodiversity Conservation in an Uncertain World*, vol. 144, no. 4, pp. 1194–1203.

Kotze, D. N., Nkosi, M. R., Riddell, E., Pollard, S. R., Ngetar, N. and Ellery, W. N. (2008) 'Evaluation of the effect of rehabilitation interventions on the health and ecosystem service delivery of the Manalana Wetland, Craigieburn Village, Mpumalanga Province', *WET-OutcomeEvaluate: An Evaluation of the Rehabilitation Outcomes at Six Wetland Sites in South Africa*, WRC Report TT 343/08, Water Research Commission, Pretoria.

Levin, S. A. (1999) *Fragile Dominion: Complexity and the Commons*, Helix Books, Cambridge, Massachusetts.

Lewanika, K. M. (2002) 'The traditional socio-economic systems for monitoring wetlands and wetland natural resources utilization and conservation: The case of the Barotseland, Zambia', Conference on Environmental Monitoring of Tropical and Subtropical Wetlands, Maun, Botswana.

McCarthy, T. S. and Hancox, P. J. (2000) 'Wetlands', in T. C. Partridge and R. R. Maud (eds) *The Cenozoic of Southern Africa*, Oxford Monograms on Geology and Geophysics, Oxford University Press, Oxford.

McCartney, M. (2001) 'Understanding dambo hydrology: Implications for development and management (Chapter 13)'. *Proceedings of an FAO Sub-regional Workshop in Harare, Zimbabwe*, 19–23 November 2001.

Meinzen-Dick, R. and Pradhan, R. (2002) 'Legal pluralism and dynamic property rights', *CAPRI Working Paper 22, CGIAR System-Wide Program on Collective Action and Property Rights*, IFPRI, Washington, DC.

Millennium Ecosystem Assessment (2003) *Ecosystems and Human Well-being: A Framework for Assessment*, Island Press, Washington, DC.

Ostrom, E. (1990) *Governing the Commons: The Evolution of Institutions for Collective Action*, Cambridge University Press, Cambridge.

Pirot, J., Meynell, P. and Elder, D. (eds) (2000) *Ecosystem Management: Lessons from Around the World. A Guide for Development and Conservation Practitioners*, IUCN, Gland and Cambridge.

Pollard, S. R. and Cousins, T. (2008) 'Towards integrating community-based governance of water resources with the statutory frameworks for Integrated Water Resources Management: A review of community–based governance of freshwater resources in four southern African countries to inform governance arrangements of communal wetlands', WRC report TT.328/08, Water Research Commission, Pretoria.

Pollard, S. R. and Cousins, T. (2010) 'Towards the wise use of wetlands: A draft framework for the assessment of adaptive capacity and field testing in Mutale, Venda', final report to Working for Wetlands Programme and Mondi Wetlands Programme, December 2010.

Pollard, S. R. and du Toit, D. (2007) 'Guidelines for strategic adaptive management: Experiences from managing the rivers of the Kruger National Park, South Africa', Ecosystems, Protected Areas and People, Project Planning and managing protected areas for global change, IUCN/UNEP/GEF project GF/2713-03-4679.

Pollard, S. R. and du Toit, D. (2011) 'Towards adaptive integrated water resources management in southern Africa: The role of self-organisation and multi-scale feedbacks for learning and responsiveness in the Letaba and Crocodile catchments', *Water Resources Management*, vol. 25, no. 15, pp. 4019–4035.

Pollard, S. R., Biggs, H. and du Toit, D. D. (2008a) Towards a socio-ecological systems view of the Sand River Catchment, South Africa: An exploratory resilience analysis, Water Research Commission, project K8/591, Pretoria.

Pollard, S. R., Kotze, D. and Ferrrari, G. (2008b) 'Valuation of the livelihood benefits of structural rehabilitation interventions in the Manalana Wetland', in W. Ellery and D. Kotze (eds) *WET-OutcomeEvaluate: An Evaluation of the Rehabilitation Outcomes at Six Wetland Sites in South Africa*, WRC Report TT 343/08, Water Research Commission, Pretoria.

Pollard, S. R., Perret, S., Kotze, D., Lorentz, S., Riddell, E. and Ellery, W. (2008c) 'Investigating interactions between biophysical functioning, usage patterns, and livelihoods in a wetland agro-ecosystem of the Sand River Catchment through dynamic modelling', Challenge Program on Water and Food's Project 30 – Wetlands-based livelihoods in the Limpopo Basin, Association for Water and Rural Development (AWARD), Acornhoek, South Africa.

Pollard, S. R., Shackleton, C. and Carruthers, J. (2003) 'Beyond the fence: People and the Lowveld landscape', in J. T. D. Toit, K. H. Rogers and H. C. Biggs (eds) *The Kruger Experience: Ecology and Management of Savanna Heterogeneity*, Island Press, Washington, DC.

Pollard, S. R., du Toit, D., Cousins, T., Kotze, D., Riddell, E., Davis, C., Addey, S. and Mkhabela, B. (2009) 'Sustainability indicators in communal wetlands and their catchments: Lessons from Craigieburn wetland, Mpumalanga, Pretoria', Water Research Commission, Pretoria.

Pollard, S. R., Kotze, D., Ellery, W., Cousins, T., Monareng, J. and Jewitt, G. (2005) 'Linking water and livelihoods: The development of an integrated wetland rehabilitation plan in the communal areas of the Sand River Catchment as a test case', Association for Water and Rural Development, South Africa, www.award. org.za/File_uploads/File/WL%20Phase%20I%20final%20report%202005-2. pdf (accessed 17 October 2012).

Pollard, S. R., Perez de Mendiguren, J. C., Joubert, A., Shackleton, C. M., Walker, P., Poulter, T. and White, M. (1998) *Save the Sand Phase 1 Feasibility Study: The Development of a Proposal for a Catchment Plan for the Sand River Catchment*, Deptartment of Water Affairs & Forestry, Pretoria.

Ramsar (2007) 'A conceptual framework for the wise use of wetlands', Ramsar handbook for the wise use of wetlands, 3rd edition, vol. 1, Ramsar Convention Secretariat, Gland.

Republic of South Africa (RSA) (1998) *The National Water Act, Act 36 of 1998*, Republic of South Africa Government Gazette, no. 19182, Cape Town.

Riddell, E. S., Lorentz, S. A. and Kotze, D. C. (2012) 'The hydrodynamic response of a semi-arid headwater wetland to technical rehabilitation interventions', *Water SA*, vol. 38, no. 1, pp. 55–66.

Schulze, R. (1989) 'ACRU: Background, concepts and theory', Report 35, Agricultural Catchments Research Unit, Department of Agricultural Engineering, University of Natal, Pietermaritzburg.

Shackleton, S., Campbell, B., Lotz-Sisitka, H. and Shackleton, C. (2008) 'Links between the local trade in natural products, livelihoods and poverty alleviation in a semi-arid region of South Africa', *World Development*, vol. 36, pp. 505–526.

Snowden, D. J. (2003) 'Managing for serendipity or why we should lay off "best practice" in knowledge management', *Journal of Knowledge Management*, ARK May 2003.

Stibbe, A. (2004) 'Environmental education across cultures: Beyond the discourse of shallow environmentalism', *Language and Intercultural Communication*, vol. 4, no. 4, pp. 242–260.

Von Bertalanffy, L. (1968) *General System Theory: Foundations, Development, Applications*, G Braziller, New York.

Von der Heyden, C. J. and New, M. G. (2003) 'The role of the dambo in the hydrology of a catchment and the river network downstream', *Hydrology and Earth System Sciences*, vol. 7, no. 3, pp. 339–357.

Walker, B. and Salt, D. (2006) *Resilience Thinking: Sustaining Ecosystems and People in a Changing World*, Island Press, Washington, DC.

Walker, B., Holling, C. S., Carpenter, S. R. and Kinzig, A. (2004) 'Resilience, adaptability and transformability in social ecological systems', *Ecology and Society*, vol. 99, no. 2, art. 5.

Wood, A. and van Halsema, G. E. (2008) *Scoping Agriculture – Wetland Interactions: Towards a Sustainable Multiple-response Strategy*, FAO, Rome.

6 Assessing the ecological sustainability of wetland cultivation

Experiences from Zambia and Malawi

Donovan Kotze

Summary

This chapter reports on an assessment of the ecological condition and sustainability of cultivation of the Chikakala dambo, Zambia and the Katema dambo, Malawi using the WET-SustainableUse framework, which relies on rapidly described indicators of wetland functioning. Cultivation is confined mainly to the dry season (particularly in Katema) and cultivation in the core, wettest areas of the wetland is avoided, which contributes to reducing the impacts of cultivation. However, plot-level practices at both sites (e.g. inadequate use of soil-building crops and a high level of tillage) contribute negatively to ecological condition by reducing soil organic matter (SOM) levels and nutrient retention and replenishment. In addition, at Chikakala, turf burning further adds to the impacts within plots. However, turf burning has, as yet, not resulted in any obvious major wetland degradation, and the cultivated plots occupy only 5 per cent of the overall wetland, allowing 'space' for the fallowing and recovery of plots. Thus, the multiple demands and multiple sustainable benefits appear to be well accommodated, and cultivation appears to be sustainable provided that the overall extent of cultivation does not increase. In contrast, inadequate space is available at Katema for fallowing, where cultivated plots occupy 40 per cent of the overall wetland. Thus, while the multiple demands are currently being reasonably well accommodated, in the long term it will become increasingly difficult to do so at the overall wetland level, and sustainability is likely to be compromised.

Introduction

Dambos, seasonally waterlogged wetlands, usually treeless, at or near the head of a drainage network, are widespread across Africa's tropical savannas (Whitlow, 1989). The direct use of dambos by subsistence farmers has long played a critical role in the livelihoods of local populations (Acres *et al.*, 1985). At the same time, however, limitations and environmental risks associated with dambo cultivation are also recognized, particularly in the light of rapidly increasing human populations, and the human pressures on dambos (Acres *et al.*, 1985; Whitlow, 1985).

The Chikakala dambo in northern Zambia and the Katema dambo in central Malawi (Figure 6.1) are examples where there is a strong link between dambos and rural livelihoods (Mbewe, 2007; Msukwa, 2007). Local livelihoods are also strongly connected to the miombo woodland which surrounds both dambos. Miombo woodland has nutrient-poor soils and more than 95 per cent of the rain falls in a single 5–7 month wet season (Campbell *et al.*, 1996), with profound consequences for local livelihoods and the role that dambos play in these livelihoods. Nutrients are quickly depleted under continuous cultivation of upland fields, and therefore shifting cultivation has been traditionally practised. However, with increasing human populations, available woodland has become increasingly limited. This has resulted in dambos, with their generally higher nutrient levels, becoming increasingly attractive for cultivation. Further adding to the attractiveness of dambos is their residual moisture which allows cultivation to be extended through the dry season.

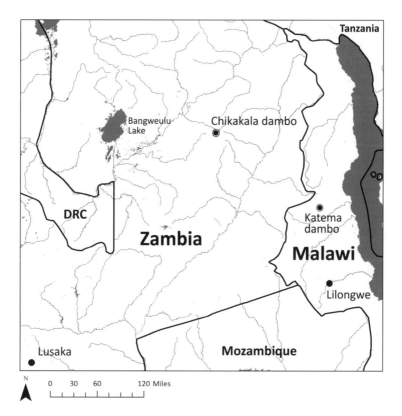

Figure 6.1 Location of Chikakala and Katema dambos

The ecosystem services (ESS) and livelihood benefits supplied by dambos are underpinned by the ecological condition of the dambo, which needs to be assessed if sustainability of use is to be better understood. Although sustainable use principles (e.g. Mharapara, 1994) and guidelines for the classification of the agricultural potential of wetlands (e.g. McCartney *et al.*, 2005) already exist, these lack specific criteria for assessment of the ecological condition of wetlands. Thus, a framework, referred to as WET-SustainableUse (Kotze, 2010) was developed, based closely on the framework of Macfarlane *et al.* (2009). WET-SustainableUse provides a set of indicators for the assessment of each of five components of wetland ecological condition, namely hydrology, geomorphology, SOM accumulation, nutrient cycling and retention, and vegetation composition. The indicators, which provide the basis to characterize human pressure on wetlands and to assess the sustainability of use, can be described rapidly. Consequently, the framework has relevance to countries that are poor in data and resources.

This chapter reports on an assessment of the ecological condition and sustainability of cultivation in the Chikakala and Katema dambos using WET-SustainableUse (Kotze, 2010). Based on the results, recommendations are provided in terms of reconciling some of the tensions between conservation and agro-development and how management initiatives can better contribute to the sustainability of wetland cultivation. The assessment draws on data collected in 2008 from a portion of the Katema dambo (Kotze *et al.*, 2008) and from 2007 to 2009 at Chikakala (Nyirenda, 2008; Sampa, 2008; Kotze, 2009; and unpublished soil fertility status data collected in 2009). The focus of the sustainability assessment was on the impact of cultivation, with particular emphasis given to the turf ridging and burning practices in Chikakala. Although sustainability is widely recognized to encompass social and economic dimensions, it is important to note that the assessment concentrates on ecological sustainability only.

The framework used to carry out the assessment

WET-SustainableUse is a modular approach for assessing the present ecological condition of wetlands. It attempts to account for some of the key interacting processes that take place within a wetland and synthesize this information by evaluating inter-related components of ecological condition. WET-SustainableUse assists in identifying the likely contribution of specific uses to discrete components of ecological condition (Table 6.1).

For each of the five components, the impacts of human activities on ecological condition were scored based on readily observed indicators (e.g. artificial drainage channels). The spatial *extent* of the impact of individual activities was assessed and then the *intensity* of impact of each activity in the affected area was evaluated based on a scale of 0 (no impact) to 10 (critical impact) (Table 6.2). The extent and intensity were then combined to

Table 6.1 Key components considered in assessing the extent to which use alters the ecological condition of a wetland

Key components of ecological condition	Rationale
Hydrology (water distribution and retention). This can be altered through: 1) changes in water inputs as a result of human activities in the catchment upstream of the wetland; or 2) modifications within the wetland (notably those resulting from the excavation of artificial drainage channels) that alter the water distribution and retention patterns within the wetland. Both these types of alterations are evaluated and combined into a measure of the hydrological condition of the wetland.	Hydrology is the primary determinant of wetland functioning. The hydrological conditions in a wetland affect many abiotic factors, including soil anaerobiosis (waterlogging), availability of nutrients and other solutes, and sediment fluxes (Mitsch and Gosselink, 1986). These factors, in turn, strongly affect the fauna and flora that develop in a wetland.
Geomorphology (sediment trapping/erosion). This is defined as the distribution and retention patterns of sediment within the wetland, including both mineral and organic sediments. This component focuses on evaluating changes in erosional and depositional patterns within the wetland as a result of human activities.	Wetlands are generally net accumulators of sediment (Ellery *et al.,* 2008), which affects the landform of the wetland, which, in turn, has a feedback effect on hydrology. Sediment retention is also important for maintaining the wetland's on-site agricultural productivity as well as being potentially important for downstream water users by enhancing nutrient retention.
SOM accumulation/depletion. SOM accumulates in the upper soil layers, and in wetlands is typically more abundant than in non-wetland areas as a result of waterlogging – slowing the rate of organic matter decomposition. This component assesses, based on indicators, the extent to which human activities disrupt SOM accumulation.	SOM makes a significant contribution to wetland functioning and productivity (e.g. enhanced water holding capacity of the soil and enhanced cation exchange capacity (CEC) of the soil (Millar and Gardiner, 1998; Mills and Fey, 2003; Sahrawat, 2004).
The retention and internal cycling/leakage of nutrients. It is recognized that this is very closely linked with the retention of sediment, as well as being affected directly by the presence and growth of vegetation, another feature affected by human activities (e.g. crop production).	Wetlands are generally effective in retaining and cycling nutrients, which is important for maintaining the wetland's on-site productivity in terms of growth of natural vegetation and crops as well as being potentially important for downstream water users by enhancing nutrient retention (Mitsch and Gosselink, 1986).
The natural composition of the wetland vegetation. The natural composition of a wetland is generally most directly and dramatically altered by the clearing of vegetation for cultivation.	The particular composition of wetland vegetation is of significance in itself for biodiversity and provides habitat for a range of fauna. Particular plant species may also have direct economic importance (e.g. for use in craft production).

Table 6.2 Guideline for assessing the intensity of impact on wetland ecological
         condition

Impact intensity score	Description
0–0.9	Unmodified. No change in ecosystem processes is discernible.
1–1.9	Largely natural with few modifications. A slight change in ecosystem processes is discernible.
2–3.9	Moderately modified. A moderate change in ecosystem processes has taken place but the ecosystem remains predominantly intact.
4–5.9	Largely modified. A large change in ecosystem processes has occurred.
6–7.9	The change in ecosystem processes is great but some remaining features are still recognizable.
8–10	Modifications have reached a critical level and the ecosystem processes have been modified completely.

Source: modified from Macfarlane *et al.*, 2009

determine an overall *magnitude* of impact also on a scale of 0 to 10, as follows: **Magnitude = Extent/100 × Intensity**. For example, if a given activity
was affecting 25 per cent of the wetland and its intensity was 4 then the
magnitude of the impact is $25/100 \times 4 = 1.0$.

Although the components given in Table 6.1 are assessed separately,
it is acknowledged that they are closely interlinked. Water distribution
and retention has the most influence on the other factors and is the least
affected by the other factors, and can therefore be considered as the primary driving process. Conversely, nutrient cycling is the most affected by
the other factors and has the least influence on the other factors. The main
influence of nutrient cycling is through plant growth and the organic matter accumulation that results from this growth, i.e. its influence on the
other factors is indirect through organic matter accumulation.

Following the assessment of ecological condition, the implications of
this for the delivery of ESS is discussed with reference to a few key selected
regulatory/supporting services, notably the assimilation of nutrients and
biocides, the storage of water and the maintenance of soil for sustained
cultivation.

An overview of the two dambo sites and their cultivation

Two dambos were selected for the study (Figure 6.1) with inherently similar biophysical features, but with contrasting levels of use and catchment
contexts. The Katema dambo (12°43.368'S; 33°35.940'E) is located near
Kasungu town in the drainage basin of the Dwangwa River, which flows into
Lake Malawi, an aquatic system of regional and global importance (Thieme

et al., 2005). Chikakala dambo (11°24.222'S; 31°15.269'E) is located near Mpika town in the upper Congo River drainage basin in the catchment of the Chambeshi River, which feeds the Bangweulu Lake. Although the mean annual precipitation is higher at Chikakala than Katema (~1,100 mm compared with ~800 mm) both sites have an extended dry season from May to September/November.

Chikakala dambo (754 ha) is bigger than Katema dambo (a 34 ha arm within an overall wetland of 408 ha) and occupies a larger proportion of its catchment (29 per cent compared to 13 per cent). However, both wetlands have a similar hydrogeomorphic setting, occurring near the head of the valley and comprising a valley bottom and extensive hillslope seepages. In both wetlands the margins are dominated by grasses, notably *Hyparrhenia* spp., while sedges dominate the wetter body of the wetland, including *Rhyncospora* spp., *Fuirena* spp. and *Eleocharis* spp.

Traditionally, the predominant agriculture practised in the uplands around Chikakala dambo was *chitemene* cultivation, where the branches and foliage of trees are collected from an area of several hectares (the 'outfield'), and gathered in the 'infield', an area about 0.4 ha in size, allowed to dry and then burnt just before the rains. When the infield is exhausted after a few years of cultivation, it is abandoned and traditionally left fallow for 20 to 30 years (Strømgaard, 1984a and b). *Chitemene* cultivation allows for nutrients from across a wide area to be collected and concentrated in the infield through the medium of the ash. This assists in dealing with the inherently nutrient-poor soils of the miombo woodlands (Strømgaard, 1984a and b). In the Katema area, similar shifting cultivation is likely to have been carried out in the past, but now, like in most of Malawi, the demand for land is so high that fallow periods are much shorter, and most of the cultivation is effectively permanent.

Both dambos have probably been used since pre-colonial times, but the extent of use increased noticeably in the last two decades. At Chikakala this was stimulated by the 1990/1991 drought, together with fertilizer shortages, which caused upland harvests to fail (Sampa, 2008), while in Katema it was stimulated by a series of droughts from 2000 (Wood, 2005). The residual moisture in both dambos supports food production in the dry season and this continues into the rains in Chikakala. For subsistence farmers relying on upland cultivation, the early wet season is generally the most intense 'hunger period' because most/all of the food from the previous wet season has been consumed (particularly during a drought) and the harvest is not yet ready from the current rain-fed crops (Kotze *et al.*, 2008; Sampa, 2008). Crops grown in the Chikakala dambo include early maturing varieties of pumpkin, squash and maize, as well as other crops such as tomatoes, onions, cabbage and beans. (Sampa, 2008). The main crops grown in Katema dambo are maize, sugar cane, rape, mustard, green beans, tomatoes, Irish potatoes and bananas. Crops are used for direct household consumption as well as for sale (Mbewe, 2007; Msukwa, 2007; Kotze *et al.*, 2008; Sampa, 2008). The total number of households using the wetlands was 28 for Katema (Msukwa, 2007) and 67 for Chikakala (Mbewe, 2007). The level of dependency of these households on the wetlands was high. Katema dambo contributed 37 per cent of the food consumed by the household. Furthermore, levels of local poverty are very high, with more than 70 per cent of the households categorized as poor and subject to at least seasonal shortages of food (Msukwa, 2007). Although not quantified, Chikakala is likely to have similar levels of dependency, with 87 per cent of households classified as poor (Mbewe, 2007). At both sites, the income from dambo crop sales makes an important contribution to basic household needs, e.g. for cooking oil and clothes. Chikakala in particular is fairly market orientated, with links to mining towns over 400 km away.

The traditional land preparation at Chikakala dambo involved digging up the surface soil together with plant roots in thick 'slices' of turf. After being left to dry, the turf was burnt, and then cultivation took place in the ash-rich soils (Sampa, 2008). However, burning of the thick-cut turf often proved incomplete, leading to variable yields. Thus, during the 1990s a refined method was developed. This involved cutting thin turfs, drying them on the ground (grass side down), gathering the turfs into well-ventilated ridges, burning these ridges and re-ridging after burning (which covers the ash with soil, thereby preventing the ash from being blown away) (Sampa, 2008). 'Overall this improves nutrient availability and moisture retention ... this has led to a method which can sustain three to four harvests in succession over two years without chemical fertilizers and without major water application, if a moist site in a dambo is selected' (Sampa, 2008: 3).

Soil analyses in Chikakala confirm that available phosphorus is significantly higher in the cultivated ridges than in the uncultivated areas (26.7 ± 3.5 ppm [n=3] compared with 10.0 ± 9.6 ppm [n=3]), whereas nitrogen appears lower (0.10 ± 0.02 per cent [n=3] compared with 0.37 ± 0.46 per cent [n=3]). The turf burning appears to concentrate phosphorus in several ways. First, phosphorus that is bound in the combusted organic matter is not lost through volatilization, as occurs with much of the nitrogen, and instead remains in the ash. Second, much of the SOM and available nutrients are naturally concentrated in the topsoil, which is gathered together in the ridges, further concentrating available nutrients. In addition, the direct heating of the soil and increased pH that results from the ash causes some of the unavailable soil phosphorus to be converted into a form available to plants (Giardina *et al.*, 2000).

The turf ridging and burning cultivation method at Chikakala provides a very effective means of accessing available nutrients and residual moisture for dry season crop production, and the method appears to be 'well tuned' to the particular biophysical constraints and opportunities present in the Chikakala dambo. Nevertheless, the effectiveness of the method in providing nutrients in a relatively nutrient-poor system has important implications for replenishing nutrient reserves, as will be explored further in the results of the sustainability assessment.

At Katema, the burning of turf is not practised, but wetland plots are subject to a high level of tillage. Although some burning of weed and crop residues is practised, the extent to which residues are returned to the plot either as compost or unprocessed residues, buried in trenches within the dambo, is high.

Results of the WET-SustainabileUse assessment

Impacts on the hydrology of the wetland

Impacts on the quantity and pattern of water inputs to the wetland as a result of human activities in the wetland's upstream catchment

Although the extent of cultivation in the catchment of Katema is considerably higher than in Chikakala (Table 6.3), its effect is moderated to some extent by the fact that a buffer area immediately adjacent to the dambo remains under natural woodland vegetation, as recommended by Thawe (2008). This reduces surface runoff into the dambo and deposition of sediment there.

When assessing the effect of human activities on the input of water to the dambo, it is necessary to consider the net effect of those activities that decrease inputs together with those activities that increase inputs. In both dambos, point source effects (e.g. pumping from a borehole) are limited, and therefore have negligible effect.

Trees are consumers of water, generally more so than herbaceous plants (Dye *et al.*, 2008) owing to deeper roots and greater leaf area. Clearing miombo woodland for cultivation generally results in an increase in the amount of water delivered to a dambo from its catchment, particularly in surface flows (e.g. du Toit, 1985; Mumeka, 1986; Whitlow, 1992) which may, in turn, contribute to increased erosion (Whitlow, 1989).

Evidence of greatly increased surface runoff (e.g. as a result of hardened surfaces in the catchment) is lacking at both dambos. Further, there is very little evidence of recent sediment deposition in the dambos, which is generally indicative of increased surface flows leading to increased levels of erosion. However, given the land-cover situation described in Table 6.3,

Table 6.3 Land-cover types in the upstream catchments of the two dambos (as a percentage of the upstream catchment area)

Landcover	Chikakala	Katema
Natural/near-natural miombo woodland vegetation	39%	18%
Semi-natural recovering *chitemene* areas (includes infields, outfields and recovering areas)	48%	6%
Currently cultivated (including some recently fallow areas)	12%	66%
Homesteads, roads, tracks and paths	1%	10%

Note: approximate extent based on aerial photography and satellite imagery interpretation, with some field validation.

Chikakala and Katema are assigned a score of 1 and 2 respectively in terms of impacts resulting from altered pattern and intensity of water inputs.

Impacts on the distribution and retention of water in the wetland as a result of cultivation within the wetland

The ridges in the Chikakala plots are designed to control waterlogging (which limits the growth of most crops) while at the same time they provide adequate supply of water for crops (Sampa, 2008). This is achieved by locating the plots within areas in the dambo that remain fed by seepage waters through most of the dry season, but avoiding the excessively wet central areas of the dambo and the drier margins. The ridging system concentrates naturally diffuse flows within the furrows, which are mainly orientated diagonally down the slope. In addition, on the ridges, the land surface is raised relative to the water table by about 10 to 20 cm. Overall, therefore, the level of wetness of the ridged areas is lowered, but this is not severe given the fairly shallow nature of the ridges and furrows. In Katema, the plots are not ridged and furrowed, although there are drainage furrows around some of the cultivated plots, while raised beds are used in the early dry season when the dambo has a high water table. Thus, the intensity of impact is lower in the Katema plots than in the Chikakala plots (Table 6.4). However, given that the extent of the plots is much greater in Katema than Chikakala, the magnitude of impact is greatest in Katema, but is nevertheless still not high (Table 6.4).

In both of the dambos, once the plots are abandoned, any ridges and/or furrows present are left intact. Although the ridges will gradually become flatter and sediment is likely to accumulate in the furrows, the draining effect will persist long after the cultivation has been abandoned, which is reflected in the score assigned to old abandoned plots in Table 6.4.

Table 6.4 Assessment of the current impacts of cultivation within the dambo on dambo hydrology

Land-use type	Chikakala			Katema		
	Intensity	Extent	Magnitude	Intensity	Extent	Magnitude
Currently cultivated and recently abandoned	4	5%	0.2	3	40%	1.2
Old (long) abandoned	2	10%	0.2	1	13%	0.1
Natural/near-natural	0	85%	0	0	47%	0
Overall			0.4			1.3

Note: intensity score refers to the intensity of impacts on a scale of 0 (no impact) to 10 (critical impact). The rationale for the intensity scores is given in the text accompanying the table.

Note: recently abandoned refers to land which appears to have been cultivated in the previous dry season, while old (long) abandoned refers to lands that were previously cultivated but which have been left uncultivated for more than a single growing season; total magnitude of change is on a scale of 0 (pristine) to 10 (complete loss of native species).

Impacts on the retention (or erosion) of sediment in the wetland as a result of cultivation within the dambo

Several activities take place in the cultivated plots in the two dambos that contribute to an increased vulnerability to erosion:

- A high level of soil tillage – disrupts soil structure and destroys plant roots that contribute to the strength of the soil.
- Combustion of the soil (in the case of Chikakala) – makes the soils more friable, which has a similar impact to the above.
- Diminished SOM levels due to cultivation – less SOM reduces the physical strength of soils, especially in sandy to loamy soils.
- Concentration of surface water flow – facilitates soil erosion.
- Diminished soil cover – reduces protection of the soil against rainsplash erosion and water flow erosion (to a lesser degree).

However, several factors act to limit the actual erosion that has taken place, including the following:

- Cultivation takes place mainly outside those portions of the dambo that appear to be most susceptible, i.e. the lowest-lying areas of the dambo that carry most of the longitudinal water flows, as recommended by Thawe (2008).
- The time when the soils are most vulnerable (i.e. when recently tilled) generally coincides with the dry season, when the threat of erosion is lowest, but where a crop is grown into the early wet season then removal of the crop may expose the soil to water erosion.
- The inherent erosion hazard of the dambos and their upstream catchments is low given the gentle slope of both (Table 6.5).

Table 6.5 Assessment of the current impacts of cultivation within the dambo on the retention of sediment/erosion

Land-use type	Chikakala			Katema		
	Intensity	Extent	Magnitude	Intensity	Extent	Magnitude
Currently cultivated and recently abandoned	2	5%	0.1	1	40%	0.4
Old (long) abandoned	1	10%	0.1	0	13%	0
Natural/near-natural	0	85%	0	0	47%	0
Overall			0.2			0.4

No direct evidence of erosion was observed in either of the dambos. Thus, the intensity of impact was assigned a relatively low score in both dambos, but slightly higher in Chikakala than Katema primarily owing to the combustion of soil in Chikakala.

Impacts on the accumulation of soil organic matter in the wetland as a result of cultivation within the wetland

SOM makes a significant contribution to wetland functioning and productivity (e.g. by enhancing soil water holding capacity), and can be profoundly affected by different land-use practices. Several factors are likely to contribute to diminished SOM levels in the cultivated plots (Table 6.6). The

Table 6.6 Factors contributing to likely diminished SOM levels in the cultivated plots

Factors	Rationale
Tillage of the soil, which is high for both dambos. It occurs for preparing the beds and ridges, and for planting and weeding.	Tillage increases the oxygen levels in the soil, which increases the rate of SOM decomposition. Another key mechanism by which the rate of decomposition of organic matter is reduced is through the organic matter being physically protected within soil aggregates, which are broken by tillage (Six *et al.*, 2002).
Reduced cover of the soil, which occurs mainly when the beds are being prepared and in the first few weeks following preparation of the ridges before the crops have developed good aerial cover.	The greater the exposure of the soil, the greater the extent to which the soil is subject to temperature fluctuations, which in turn contribute to increased levels of SOM depletion (Six *et al.*, 2002).
Combustion of the soil (only found in Chikaklala).	If the soil material is well combusted then the amount of organic material potentially lost is very high in the short term. However, SOM levels on the burnt-turf ridges do not appear to be significantly lower than those in uncultivated lands which suggests that it is the dried-out plant material rather than the organic component of the soil that is combusted. However, in the long term the combusted plant material would reduce the amount of organic matter potentially available for incorporation into the SOM pool.
Reduced level of wetness, exposing the soil to higher oxygen levels, and therefore more rapid decomposition rates.	Prolonged soil saturation results in the development of anaerobic soil conditions and impeding SOM decomposition (Tiner and Veneman, 1988). Therefore, the greater the level of artificial drainage, the greater the potential loss of SOM previously accumulated under the wetter conditions.
Removal of most of the plant residues from the croplands.	The amount of organic matter that would otherwise be returned to the soil is reduced.

Table 6.7 Assessment of the current impacts of cultivation within the wetland on the accumulation of SOM

Land-use type	Chikakala			Katema		
	Intensity	Extent	Magnitude	Intensity	Extent	Magnitude
Currently cultivated and recently abandoned	6	5%	0.3	5	40%	2.0
Old (long) abandoned	3	10%	0.3	2	13%	0.3
Natural/near-natural	1	85%	0.9	1	47%	0.5
Overall			1.5			2.8

returning of crop residues to the plots (which occurs particularly at Katema) contributes positively to SOM, but this in itself is unlikely to counter all the negative factors (Table 6.6).

Considering the factors given above, the severity of SOM depletion in the currently cultivated plots is assigned a higher score in Chikakala than Katema (Table 6.7) owing mainly to the high incidence of soil combustion at Chikakala and the limited burning of soil/plant residues at Katema. The impact in old abandoned plots is scored lower than in the currently cultivated areas (Table 6.7) given the absence of tillage, higher cover of the soil and cessation of soil burning, but considering the fact that the recovery of SOM, even in wetlands, is generally slow (Six *et al.*, 2002; Walters *et al.*, 2006) it is assumed not to have fully recovered.

Impacts on nutrient cycling in the wetland as a result of cultivation within the wetland

Impacts on nutrient cycling in the currently cultivated plots are assigned intensity scores in Table 6.8 given the following key factors. Nutrient cycling is compromised by a decline in SOM, and consequent decline in CEC, which affects the amount of nutrients held in the soil. The vegetation growth in cultivated plots is interrupted following the harvesting of one crop and while the next crop is still becoming established. During this 'interruption period', very little uptake of freely available nutrients by plants is taking place in the plot, and therefore these nutrients are vulnerable to being lost from the plot through leaching. Therefore, the greater the level of interruption of actively growing vegetation (between harvesting/ senescence of one crop and the full establishment of the next crop), the lower will be the capacity of plants to take up mobile nutrients and prevent them from being leached (Randall and Goss, 2001).

Perennial vegetation rapidly establishes on abandoned plots, which contributes to the retention of nutrients, and probably also to the gradual recovery

Table 6.8 Assessment of the current impacts of cultivation within the wetland on the retention of nutrients

Land-use type	Chikakala			Katema		
	Intensity	Extent	Magnitude	Intensity	Extent	Magnitude
Currently cultivated and recently abandoned	5	5%	0.3	5	40%	2
Old (long) abandoned	2	10%	0.2	2	13%	0.3
Natural/ near-natural	0	85%	0	0	47%	0
Overall			0.5			2.3

of nutrient levels. Based upon this consideration, and assuming that erosion in the abandoned plots is very limited, impacts on nutrient cycling in the long abandoned plots are assigned a low impact intensity score in Table 6.8.

The low nutrient levels in the leached acidic soils of both dambos means that *permanent* cultivation of the dambos is unlikely to be sustained without additions of manure or inorganic fertilizers. At both sites, manure is in short supply as there is little livestock, and farmers are generally not able to afford inorganic fertilizers. Therefore, careful management of nutrients, including fallow periods, is required for sustained production. Recovery rates of soil nutrients in leached acidic soils are typically low (Sanchez, 1976; Williams *et al.*, 2008). Both Wood (2009, pers. comm., Centre for Wetlands, Environment and Livelihoods, University of Huddersfield) and Strømgaard (1984a) report fallow periods of 20 to 30 years being required in *chitemene* cultivation in the upland areas of Mpika and nearby areas. Given that the dambos appear to be more inherently fertile than the uplands, the fallow period would probably not need to be as long. Sahrawat (2004) also notes that when compared with upland soils, wetland ones are better endowed in maintaining fertility, possibly owing to their generally higher inherent fertility. However, given the low nutrient status of the dambo soils, the fallow period required is unlikely to be considerably less than that required for uplands and is likely to be at least ten years given that the dambos are not subject to high sediment deposition levels, which could otherwise potentially replenish depleted nutrient reserves in the dambo. (However, such erosion from the uplands would probably be associated with negative impacts on the hydrology of the wetland.) In addition, cultivation results in the lowering of SOM levels, which are likely to recover slowly.

The smaller the sustained extent of a given wetland that is cultivated, the longer will be the potential fallow period relative to the actively cultivated period. For example, in the case of Chikakala, 5 per cent of the area of a wetland is cultivated or very recently cultivated, which allows for 19 years of fallow for every year of cultivation. Given the preceding discussion, such a period would probably be adequate even for a low-nutrient dambo such as Chikakala. In contrast, the 40 per cent currently cultivated or very recently cultivated in Katema does not allow for adequate fallow, and therefore this extent needs to be reduced together with further investment in replenishing soil fertility. Work with local farmers by the non-governmental organizations (NGOs), Wetland Action (WA) and Malawi Enterprise Zone Association (MALEZA) (e.g. as documented by Thawe, 2008) has contributed positively to promoting this investment, e.g. through composting. Even so, there is scope for greater investment, e.g. cultivating soil-building crops, particularly those such as *Sesbania sesban* (Egyptian pea) which are well adapted to the poorly drained conditions of dambos. In a trial in Malawi under farmer-managed conditions, the incorporation of *S. sesban* as a relay crop with dambo-cultivated maize had a significant positive contribution to soil fertility and maize yield (Phiri *et al.*, 1999a and b).

Impacts on vegetation

Vegetation structure has an important direct influence on hydrological flows, particularly tall, dense, robust vegetation that helps slow down the flow and protect the soils against erosion. Further, as discussed in the previous section, vegetation plays a central role in nutrient cycling and retention. The plant species composition of the vegetation also has an important influence over the condition of a wetland, as highlighted in Table 6.1, as well as influencing the provisioning services of a wetland, e.g. certain species have particular value for handcrafts. Human activity, such as cultivation, plant collection, livestock grazing and burning, can profoundly affect the vegetation structure (cover and surface roughness) and so affect water flow, nutrient cycling and erosion.

In order to assess the impacts on the vegetation as a whole and species composition in particular, four general disturbance classes can be recognized: 1) current cultivation and recent fallow; 2) older fallow; 3) areas deeply flooded; and 4) areas that have not been cultivated. For each of the disturbance classes a method for subjectively allocating an *intensity of impact score* has been developed (Table 6.9).

Data on vegetation are only available for Katema in Malawi. This shows that there has been a large impact from cultivation (Table 6.10). The consequences of this are very considerable in terms of the contribution of the dambo to maintaining biodiversity. If most of the dambos in the overall landscape were similarly impacted then the consequences of this would be regional, suggesting that some wetland areas may need to be set aside and protected from cultivation. However, in terms of the impact on the functioning of the dambo hydrologically, and for agriculture, the impacts are less severe as the natural vegetation has been maintained in the centre of the dambo where the risks of erosion would be greatest.

Table 6.9 Impact categories for assessing the intensity of impacts on vegetation integrity within disturbance classes

Impact category	Description	Intensity of impact score
None	Vegetation composition appears entirely natural.	0.5
Small	A very minor change to vegetation composition is evident at the site (e.g. abundance of ruderal, indigenous invasive slightly higher than would be the case naturally).	1.5
Moderate	Vegetation composition has been moderately altered but introduced; alien and/or increased ruderal species are still clearly less abundant than characteristic indigenous wetland species.	3
Large	Vegetation composition has been largely altered and introduced; alien and/or increased ruderal species occur in approximately equal abundance to the characteristic indigenous wetland species.	5
Serious	Vegetation composition has been substantially altered but some characteristic species remain, although the vegetation consists mainly of introduced, alien and/or ruderal species.	7
Critical	Vegetation composition has been almost totally altered, and in the worst case all indigenous vegetation has been lost (e.g. as a result of a parking lot).	9

Table 6.10 Assessment of the current impact of cultivation within dambos on the vegetation composition within different disturbance units

Land-use type	Chikakala			Katema		
	Intensity	Extent	Magnitude	Intensity	Extent	Magnitude
Currently cultivated and recently abandoned	n.a.	n.a.	n.a	9	40%	3.6
Old (long) abandoned	n.a.	n.a.	n.a.	6	13%	0.9
Natural/ near-natural	n.a	n.a.	n.a.	2	47%	0.9
Overall			n.a.			5.4

Further details of the Katema vegetation assessment are contained in Kotze *et al.* (2008) and Kotze (2011), which highlight that certain vegetation types are much less resilient to recovering their original species composition than other vegetation types.

A summary of the overall ecological condition of the two dambos and the implications of this for ecosystem service delivery

Based upon the assessments given in the previous sections, but excluding the incomplete vegetation one, a summary of the ecological condition was derived for both dambos (Table 6.11). Chikakala is in a better condition in terms of all of the components of ecological condition examined, primarily because the extent of cultivation in the dambo is considerably less than in Katema.

In order to briefly examine the implications of these ecological conditions in a wetland for the delivery of services by that wetland, some important links should be noted:

- Water distribution and retention/desiccation have direct conse-quences for the water that the wetland retains, particularly by providing a reservoir of soil moisture during the dry season, which is available for livelihood-related activities such as cultivation and livestock grazing.
- Sediment trapping/erosion and SOM accumulation/depletion are important determinants of a key supporting service – soil formation, which constitutes the foundation for the provision of food, realized through cultivation.
- Nutrient cycling is in itself a key supporting service, which in turn has profound implications for many provisioning services such as provid-ing fertile areas for cultivating food. Nutrient cycling also makes an important contribution to a regulating service – water purification.

Given these links and the specific changes in ecological condition reported in Table 6.11 (all impacts in Chikakala being small or negligible), current cultivation in Chikakala appears to be having minimal negative impacts on the delivery of supporting and provisioning services. Thus, the multi-ple demands and multiple sustainable benefits alluded to in Chapter 1 appear to be well accommodated. In the case of Katema, while the multi-ple demands are currently being reasonably well accommodated, in the long term it will become increasingly difficult to do so, particularly given that an inadequate proportion of the dambo is being left to accommodate the required fallow.

Table 6.11 Summary of the ecological condition for the two dambo sites identifying the ecological impacts for the four components assessed

Component of ecological condition	Dambo sites	
	Chikakala	Katema
Hydrology – (water distribution and retention/desiccation)	Small	Moderate
Geomorphology – (sediment trapping/erosion)	Negligible	Small
SOM accumulation	Small	Moderate
Nutrient cycling	Negligible	Moderate

Note: these are based on impact scores reported in previous sections (the classes are as follows: 0–0.9, negligible; 1–1.9, small; 2–3.9, moderate; 4–5.9, large; 6–7.9, serious; 8–10, critical).

Crucial cultivation practices common to both Chikakala and Katema strongly support sustainability of use. These include the fact that most cultivation occurs predominantly in the dry season (avoiding the season when erosion risk is highest and the need for artificial drainage is greatest). In addition, plots are located outside of the central areas of the dambo, where flows are greatest and erosion risks highest. In the face of global climate change, where extremes of drought and floods are predicted to increase, and Africa is identified as the most vulnerable continent (Dinar *et al.*, 2008), it will become increasingly important to maintain these practices. However, there are also some important differences in terms of cultivation in the two respective dambos. At a plot level, the impacts of cultivation on wetland functioning are less in the Katema dambo than in the Chikakala dambo, owing particularly to a lower level of artificial drainage and less severe depletion of SOM through turf burning. Nonetheless, there is still scope at both wetlands for greater incorporation of soil-building plants into the cropping system, increasing use of organic manure and for blocking of furrows during fallow periods, which is likely to enhance the retention of water and the accumulation of SOM and nutrients, thereby speeding up recovery during fallow periods.

Despite the impacts of cultivation at a plot level being greater in the Chikakala dambo, when considered in the context of the overall dambo and the drainage basin, dambo cultivation in Chikakala is more environmentally sustainable than in Katema. This conclusion is based primarily on the fact that the extent of cultivated plots within the overall wetland is much greater in Katema than in Chikakala. Furthermore, Katema is located in a more intensively utilized (and therefore 'harder working') drainage basin, where the demands on its regulatory services are likely to be high.

Conclusions

Application of the WET-SustainableUse framework shows that in these two dambos, in Zambia and Malawi, wetland agriculture is having a negligible to moderate impact on the environmental elements assessed. The long-term sustainable use of these areas is possible provided certain limits to the area under cultivation are kept and guidance about cultivation sites and practices are adhered to. Overall it appears that the multiple demands on these areas can be met and that it is possible to have well-managed agriculture in the wetlands, which supports livelihoods while also sustaining multiple ESS.

In reflecting on the application of the assessment framework to the two sites, the framework provided a useful means of highlighting the likely impact of specific practices on a broad spectrum of indicators reflecting wetland ecological condition. This, in turn, helped point towards practical steps that farmers can take to contribute to reconciling some of the tensions between conservation and agro-development. These steps include continuing with existing sound practices already widespread (e.g. maintaining the core of the wetland under natural vegetation) and increasing the extent of other sound practices, notably the use of soil-building crops, which is currently limited. The steps also highlighted practices that should be restricted (e.g. frequent tillage) and suggested the introduction of new practices (e.g. blocking drainage furrows in fallow lands in order to enhance recovery).

At the same time, however, it is recognized that the framework is based primarily on readily described structural indicators of processes, rather than on direct description of the processes themselves. Thus, the framework is not seen as a substitute for specific process-based research within African wetlands, for which there is a great need, and which would assist in validating and refining the rapid assessments described in this chapter. The framework used to assess cultivation also has specific components for assessing other types of use, notably grazing and the harvesting of wetland sedges and reeds, and in its development has been applied to wetlands across a diversity of hydrogeomorphic and climatic settings (Kotze, 2010). It is important to recognize that the framework, of necessity, involves an element of subjective judgment when scoring the indicators. However, the indicators and criteria for scoring are described in detail in Macfarlane *et al.* (2009) and Kotze (2010), which are accessible on the internet to any assessor wishing to reproduce the methods (http://www.wrc.org.za/Pages/ KnowledgeHub.aspx). The explicit rationale given in the framework also enhances the potential to refine and adapt the framework where required. Thus, although further field testing is required, it is suggested that the framework could be applied more widely in Africa by NGOs and government departments operating with resource limitations, which seldom allow for the direct description of ecological processes.

An important limitation in terms of the wider application of this framework is that it is designed for application by practitioners/extension workers with technical training. Therefore, it is suggested that the accessibility of Kotze (2010) and Macfarlane *et al.* (2009) be broadened, e.g. by exploring the use of appropriate heuristics (e.g. Pollard *et al.*, 2009; Chapter 5). In this way, this framework could be revised into a form that could be used directly by local communities and the field staff of government and NGO agencies. Devolving monitoring in this way would empower these communities and field staff in the management of wetlands, help develop local governance and institutions, and so build social sustainability (see Chapters 1, 3 and 4).

It is recognized that this investigation focuses on the ecological dimension of environmental sustainability but, recognizing that ecosystem use (and protection) is by definition a social and political process (Brechin *et al.*, 2002), wetlands that do not have sustainable governance and management systems will be vulnerable to unsustainable use. Therefore it is important that the biophysically focused frameworks applied in this study be complemented strongly by an understanding of the dynamics of governance and the socio-economic factors that affect wetland use, as outlined by Kambewa (2005), Pollard and Cousins (2007) and in Chapters 3, 4 and 5 of this volume.

Finally, one further development of the framework could be considered for the future. While the demands of local communities for all of the provisioning, regulating and supporting services mentioned in this chapter lead to a focus on the dambos themselves, it is suggested that this method should be widened to incorporate a catchment-wide view of the roles of dambos in terms of upstream–downstream linkages in flow regulation and water purification.[1]

Note

1   Water purification is of greater importance in Katema than Chikakala, given that the Dwangwa basin (in which Katema is located) is more densely populated and intensively used than the Chambeshi drainage basin (in which Chikakala is located). In particular, the Dwangwa basin feeds Lake Malawi, which is buffered at the inflow by an undisturbed wetland of less than 2 km wide, compared with the Chambeshi drainage basin which feeds Bangweulu Lake, buffered by an undisturbed wetland of more than 70 km wide. Water-quality impacts have already been detected in Lake Malawi, and the Dwangwa is one of the most degraded drainage basins feeding the lake (Thieme *et al.*, 2005). It is recognized that the Katema wetland is too small and far away from Lake Malawi for land-use impacts in the wetland to be reflected in a significant change in the water quality of the lake. However, there are many similar such wetlands in the catchment of the lake under high human pressure, and the cumulative effect of all of these transformations could have potentially significant consequences for water quality in the lake, and the many livelihoods depending on the lake. Nevertheless, it should be noted that any water purification services delivered by wetlands in the drainage basin may be largely negated by intense agriculture in the Dwangwa delta.

References

Acres, B. D., Rains, A. B., King, R. B., Lawton, R. M., Mitchell, A. J. B. and Rackham, L. J. (1985) 'African dambos: their distribution, characteristics and use', in M. F. Thomas and A. S. Goudie (eds) 'Dambos: small channelless valleys in the tropics', *Zeitschrift für Geomorphologie*, Supplement 52, pp. 147–169.

Brechin, S. R., Wilshusen, P. R., Fortwangler, C. L. and West, P. C. (2002) 'Beyond the square wheel: toward a more comprehensive understanding of biodiversity conservation as a social and political process', *Society and Natural Resources*, vol. 15, no. 1, pp. 41–64.

Campbell, B., Frost, P. and Byron, N. (1996) 'Miombo woodlands and their use: overview and key issues', in B. Campbell (ed.) *The miombo in transition: woodlands and welfare in Africa*, Centre for International Forestry Research, Bogor, Indonesia.

Dinar, A., Hassan, R., Mendelsohn, R. and Benhin, J. (2008) *Climate change and agriculture in Africa: impact assessment and adaptation strategies*, Earthscan, London.

du Toit, R. F. (1985) 'Soil loss, hydrological change and conservation attitudes in the Sabi catchment of Zimbabwe', *Environmental Conservation*, vol. 12, no. 2, pp. 157–166.

Dye, P. J., Gush, M. B., Everson, C. S., Jarmain, C., Clulow, A., Mengistu, M., Gledenhuys, C. J., Wise, R., Scholes, R. J., Archibald, S. and Savage, M. J. (2008) 'Water-use in relation to biomass of indigenous tree species in woodland, forest and/or plantation conditions', *Water Research Commission Report*, K5/1462/1/08, Pretoria, South Africa.

Ellery, W. N., Grenfell, M., Kotze, D. C., McCarthy, T. S., Tooth, S., Beckedahl, H., Quinn, N. and Ramsay, L. (2008) 'WET-Origin: the origin and evolution of wetlands', *Water Research Commission Report*, no. TT 334/08, Pretoria, South Africa.

Giardina, C. P., Sanford, Jr. R. L. and Døckersmith, I. C. (2000) 'Changes in soil phosphorus and nitrogen during slash-and-burn clearing of a dry tropical forest', *Soil Science Society of America Journal*, vol. 64, no. 1, pp. 399–405.

Kambewa, D. (2005) 'Access to and monopoly over wetlands in Malawi', *International workshop on 'African Water Laws: Plural Legislative Frameworks for Rural Water Management in Africa'*, 26–28 January 2005, Johannesburg, South Africa, Department of Agricultural and Applied Economics, University of Wisconsin-Madison.

Kotze, D. C. (2009) *An assessment of the ecological sustainability of the use of three dambos in the Mpika District, Zambia*, unpublished report submitted to Wetland Action and Wetland International, Wageningen, the Netherlands.

Kotze, D. C. (2010) 'WET-SustainableUse: a system for assessing the sustainability of wetland use', *Water Research Commission Report*, no. 438/09, Pretoria, South Africa.

Kotze, D. C. (2011) 'The application of a framework for assessing ecological condition and sustainability of use to three wetlands in Malawi', *Wetlands Ecology and Management*, vol. 19, no. 6, pp. 507–520.

Kotze, D. C., Walters, D. J. and Nxele, I. Z. (2008) *A baseline description of the ecological state and sustainability of use of three selected dambos in the Kasungu District, Malawi*, unpublished report submitted to Wetland Action and Wetland International, Wageningen, the Netherlands.

Macfarlane, D. M., Kotze, D. C., Ellery, W. N., Walters, D. J., Koopman, V., Goodman, P. and Goge, M. (2009) 'WET-Health: a technique for rapidly assessing wetland health', *Water Research Commission Report*, no. TT 340/09, Pretoria, South Africa.

Mbewe, A. (2007) *Maintaining seasonal wetlands and their livelihood contributions in central southern Africa: PRA report for Chimu Project Mpika District, Zambia,* unpublished report submitted to the Wetlands and Poverty Reduction Project of Wetlands International, Wageningen, the Netherlands.

McCartney, M. P., Masiyandima, M. and Houghton-Carr, H. A. (2005) *Working wetlands: classifying wetland potential for agriculture,* Research Report 90, International Water Management Institute (IWMI), Colombo, Sri Lanka.

Mharapara, I. M. (1994) 'A fundamental approach to dambo utilization', in R. Owen, K. Verbeek, J. Jackson and T. Steenhuis (eds) *Dambo farming in Zimbabwe: water management, cropping and soil potentials for smallholder farming in the wetlands,* Cornell Instutute for Food, Agriculture and Development (CIIFAD) and the University of Zimbabwe, pp. 1–9.

Millar, R. W. and Gardiner, D. T. (1998) *Soils in our environment,* 8th edn, Prentice Hall, New Jersey.

Mills, A. J. A. and Fey, M. V. (2003) 'Declining soil quality in South Africa: effects of land use on soil organic matter and surface crusting', *South African Journal of Science,* vol. 99, pp. 429–436.

Mitsch, W. J. and Gosselink, J. G. (1986) *Wetlands,* Van Nostrand Reinhold, New York.

Msukwa, C. A. P. S. (2007) *Baseline PRA report for wetland demonstration sites, Simlemba Sustainable Rural Livelihoods Project, Kasungu District, Malawi,* report submitted to the Wetlands and Poverty Reduction Project of Wetlands International, Wageningen, the Netherlands.

Mumeka, A. (1986) 'Effect of deforestation and subsistence agriculture on runoff of the Kafue headwaters, Zambia', *Hydrological Sciences Journal,* vol. 31, no. 4, pp. 543–554.

Nyirenda, M. A. (2008) *Biodiversity assessment report for three Mpika wetlands of SAB,* report submitted to Wetland Action and Wetland International, Wageningen, the Netherlands.

Phiri, A. D. K., Kanyama-Phiri, G. Y. and Snapp, S. (1999a) 'Maize and sesbania production in relay cropping at three landscape positions in Malawi', *Agroforestry Systems,* vol. 47, no. 1, pp. 153–162.

Phiri, A. D. K., Snapp, S. and Kanyama-Phiri, G. Y. (1999b) 'Soil nitrate dynamics in relation to nitrogen source and landscape position in Malawi', *Agroforestry Systems,* vol. 47, no. 1, pp. 253–262.

Pollard, S. and Cousins, T. (2007) 'Towards integrating community-based governance of water resources with the statutory frameworks for Integrated Water Resources Management: a review of community-based governance of freshwater resources in four southern African countries to inform governance arrangements of communal wetlands', *Water Research Commission Report,* no. K8614, Water Research Commission, Pretoria, South Africa.

Pollard, S., du Toit, D., Cousins, T., Kotze, D., Riddell, E., Davis, C., Adey, S., Chuma, E. and Mkhabela, B. B. (2009) 'Sustainability indicators in communal wetlands and their catchments: lessons from Craigieburn wetland, Mpumalanga', *Water Research Commission Report,* no. K5/1709, Water Research Commission, Pretoria, South Africa.

Randall, G. W. and Goss, M. J. (2001) 'Nitrate losses to surface water through subsurface, tile drainage', in R. F. Follet and J. L. Hatfield (eds) *Nitrogen in the environment: sources, problems and management,* Elsevier Science Publishers, the Netherlands.

Sahrawat, K. L. (2004) 'Organic matter accumulation in submerged soils', *Advances in Agronomy*, vol. 82, pp. 169–201.

Sampa, J. (2008) *Striking a balance: maintaining seasonal wetlands and their livelihood contributions in central southern Africa*, sustainable dambo cultivation, report submitted to Wetland Action and Wetland International, Wageningen, the Netherlands.

Sanchez, P. A. (1976) *Properties and management of soils in the tropics*, Wiley, New York.

Six, J., Feller, C., Denef, K., Ogle, S. M., de Moraes, S. and Albrecht, A. (2002) 'Soil organic matter, biota and aggregation in temperate and tropical soils – effects of no-tillage', *Agronomie*, vol. 22, nos 7–8, pp. 755–775.

Strømgaard, P. (1984a) 'Field studies of land use under chitemene shifting cultivation, Zambia', *Geografisk Tidsskrift*, vol. 84, pp. 78–85.

Strømgaard, P. (1984b) 'The immediate effect of burning and ash-fertilization', *Plant and Soil*, vol. 80, no. 3, pp. 307–320.

Thawe, P. (2008) *Maintaining seasonal wetlands and their livelihood contributions in central southern Africa, technical report 1, functional landscape approach to sustainable wetland management: integrating wetland and catchment management in Simlemba TA, Kasungu District, Malawi*, unpublished report submitted to the Wetlands and Poverty Reduction Project of Wetlands International, Wageningen, the Netherlands.

Thieme, M. L., Abell, R., Staissny, M. L. J., Skelton, P., Lehner, B., Teugls, G. G., Toham, A. K., Burgse, N. and Olson, D. (2005) *Freshwater ecoregions of Africa and Madagascar: a conservation assessment*, Island Press, Washington, DC.

Tiner, R. W. and Veneman, P. L. M. (1988) *Hydric soils of New England*, University of Massachusetts Cooperative Extension, Massachusetts.

Walters, D. J., Kotze, D. C. and O'Connor, T. G. (2006) 'Impact of land use on community organization and ecosystem functioning of wetlands in the southern Drakensberg mountains, South Africa', *Wetlands ecology and management*, vol. 14, no. 4, pp. 329–348.

Whitlow, J. R. (1985) 'Dambos in Zimbabwe: a review', *Zeitschrift für Geomorphologie*, supplement 52, pp. 147–169.

Whitlow, J. R. (1989) 'A review of dambo gullying in south-central Africa', *Zambezia*, XVI, pp. 123–150.

Whitlow, J. R. (1992) 'Gullying within wetlands in Zimbabwe: an examination of conservation history and spatial patterns', *South African Geographical Journal*, vol. 74, no. 2, pp. 54–62.

Williams, M., Ryan, C. M., Rees, R. M., Sambane, E., Fernando, J. and Grace, J. (2008) 'Carbon sequestration and biodiversity of re-growing miombo woodlands in Mozambique', *Forest ecology and management*, vol. 254, no. 2, pp. 145–155.

Wood, A. P. (2005) *Sustainable wetland management for livelihood security, Simlemba TA, Kasungu District, Malawi: an environmental and socio-economic impact and development assessment*, unpublished report, Wetland Action, Huddersfield.

7 Sustainable management of wetlands for livelihoods

Uganda's experiences and lessons

Barbara Nakangu and Robert Bagyenda

Summary

Wetlands cover approximately 11 per cent of Uganda, and are without doubt important for the country's economic development. In spite of developing one of the world's most innovative wetland policy environments, the sustainability of Uganda's wetlands is threatened by various drivers and pressures. This has called for management interventions that balance ecosystem and livelihood needs. This chapter outlines how wetland policy has evolved in Uganda, and in particular how community-based approaches have emerged as an important means of balancing ecosystem and livelihood needs. It draws on a case study of the recent Community-Based Wetlands Biodiversity conservation project (COBWEB) to illustrate the ways in which communities can play a central role in developing their own wetland management strategies. Key lessons from the project emphasize the need to adopt an appropriate governance structure as an entry point to interact with communities; and the importance of building upon existing policy frameworks to entrench sustainability. For effectiveness, change has to start with the people, whose behavioural change will largely be triggered by evidence-based demonstration and appreciation of the direct link between wetland conservation and their livelihoods. The lessons from the project feed into wider debates on decentralized, community-based wetland management approaches in the rest of Africa.

The importance of wetlands in Uganda

Wetlands in Uganda account for approximately 11 per cent (26,000 km²) of the total land area of the country (WMD/NU, 2008). They occur as seasonal or permanent landforms, and are characterized by various vegetation types that include woodland, palms and thickets, grasslands, sedges, papyrus and floating vegetation. Uganda's wetlands provide up to 37 different services and benefits, and play a critically important role in the country's economic development (Table 7.1). They directly and indirectly support 320,000 households and 2.7 million people respectively (UBOS, 2012). Moreover, as indicated by the Wetlands Management Department (WMD) (WMD

Table 7.1 Wetland values

Direct values	Indirect values	Option values	Non-use values
Production and consumption goods and services	*Ecosystem functions and services*	*Premium placed on possible future uses and applications*	*Intrinsic significance*
• Fish	• Water quality	• Pharmaceutical	• Cultural value
• Fuel wood	• Water flow	• Agricultural	• Aesthetic value
• Construction	• Water storage	• Industrial	• Heritage value
• Sand/gravel/ clay	• Water purification	• Leisure	• Bequest value
• Thatch	• Water recharge	• Water use	• Existence value
• Water	• Flood control		
• Wild foods	• Storm protection		
• Medicines	• Nutrient retention		
• Agriculture/ cultivation	• Shore stabilization		
• Transport/ recreation	• Micro-climate regulation		
• Pasture/grazing			

Source: WMD/NU, 2008

et al., 2009), many of these wetland-dependent people are the country's poor. Similarly, in many parts of the country wetland products and services are the sole source of livelihoods and the main safety net for the poorest households. Uganda's Wetlands Sector Strategic Plan (2001–2010) identifies wetlands as one of the most vital economic resources the country has, contributing hundreds of millions of US dollars per year to the Ugandan economy through their provision of ecosystem services(ESS) ranging from the supply of papyrus, cropland and fisheries to their role in water purification, groundwater recharge and water storage. Conservative economic valuation estimates put the direct annual productive value of wetlands at US$300–600 per hectare (MFPED, 2004).

A study conducted by WMD *et al.* (2009) indicated that livestock grazing occurs in 72 per cent, cultivation of food in 37 per cent, fishing in 35 per cent, and hunting in 42 per cent of the country's wetlands. Accordingly, the study recommended that wetland management should be integrated into the country's poverty reduction efforts in order to prevent families falling further into poverty, and to create new economic opportunities by boosting product diversification, and restoring or enhancing the supply of wetland products and services. The study confirms that in Uganda, agriculture, food security, livelihoods and wetlands are closely related. Certain agricultural crops thrive best in the moist, rich wetland soils, while wetlands near agricultural lands receive deposits of fertile soils from the surrounding catchment areas. At present, some subsistence farmers grow plants in the wet areas along wetland fringes, while others cultivate drained permanent wetlands, especially in south-western and eastern Uganda. In these areas wetland agriculture is characterized by the cultivation of Irish potatoes, yams, sugar cane, vegetables, paddy rice and livestock grazing. In Kabale and Kisoro Districts, for example, a network of drainage channels run across many agricultural fields to drain water from wetlands. Wetlands and agriculture still continue to be closely bound in the country, and some crops such as Irish potatoes, vegetables and paddy rice, whose production is closely related to wetland systems, find their way to distant markets in Kampala and beyond.

Threats to wetlands

In a number of cases there have been concerns that agricultural management practices in wetlands are unsustainable and have led to environmental degradation, which reduces agricultural productivity in the wetlands (Howard *et al.*, 2009). To address this, between 2000 and 2005 the then Wetlands Inspection Division (WID) prepared a range of extension materials to guide and educate current and future wetland farmers to manage wetlands and wetland edges in a more sustainable manner (WID, 2001a; WID, 2001b; WID, 2005). These outlined ways of utilizing the agricultural potential of wetlands whilst simultaneously maintaining their ecological integrity.

The intimate relationship between the poor and wetlands is further reflected in people's livelihood opportunities and vulnerabilities. Whereas a well-managed and healthy wetland provides a wide array of options that a population can use to secure their lives as well as climb out of poverty, a poorly managed one exposes them to risks and increases their vulnerability to shocks. Risks could be due to floods, over drainage and market failures, all of which reduce the flow and use of the ESS emanating from wetlands. As Uganda's population grows and the demand for economic development increases, understanding ESS in wetland systems and how they can be nurtured and sustained for the benefit of communities is imperative.

Uganda's 2008 State of Environment Report indicates a reduction in wetland coverage, attributed mostly to conversion to cropland and settlement, sand and clay mining and dumping of household waste. The report suggests three main driving forces affecting Uganda's wetlands:

• Anthropogenic pressures, arising primarily from population growth.
• Naturally occurring conditions such as climate variability.
• Structural constraints such as limitations in financial and institutional support.

(NEMA, 2008: 104)

WMD/NU (2008), however, attributes these trends to one single overriding factor: a human tendency to favour short-term gain over long-term benefits. They also note that in modern society, wetlands still tend to be regarded as wastelands, whose services are often not recognized and are taken for granted. As a result, wetlands are among the most threatened ecosystems. Consequently, wetlands will continue to be under threat with the rise in human population and decline in dryland resource yields (WID/IUCN, 2005). For the poor in particular, meeting immediate, short-term needs often takes precedence over longer-term conservation benefits.

Although policy makers in Uganda have recognized the importance of wetlands and have been developing a management policy framework since 1986, the implementation of some policies has not been rigorous and, as a result, more and more wetland areas have undergone conversion and others have been degraded. The trend is most pronounced in the country's urban areas where many wetlands have been converted to expand these centres and pave the way for construction and other developments. Estimates suggest that during the 1990s Uganda's wetlands declined by 8 per cent (NEMA, 2001) and more recent statistics indicate a further decline of wetland coverage from 13 per cent to just 10.9 per cent of Uganda's total land area (WMD *et al.*, 2009). Between 1994 and 2008 wetland cover reportedly declined from 37,575 km^2 to 26,308 km^2.

One of the key challenges for implementing Uganda's wetland policies stems from the inadequate financial and human capacities provided for their management within government. Whereas the national wetland policy devolved wetland management responsibilities to local government at the district level, this has not been supported with the necessary capacity to operationalize it (Glass, 2007); it was more about transferring secure rights and power to local governments. It is not uncommon to find local governments without wetland officers and, where they are present, the budgets available are too limited to support any operational costs. Overall, the financing trend for the environment and natural resources sector indicates a decline from Uganda shillings 88,288 million (US$23.2 million) during the financial year 2003/2004 to just 34,444 million (US$15.3 million) during the financial year 2008/2009 (MWLE, 2009).

The limited capacity of local governments to undertake their devolved wetlands roles coupled with a high population growth rate of 3.3 per cent (UBOS, 2012) and a fast-growing economy has put considerable pressure on natural resources, especially wetlands. This is, however, happening in a vacuum without studies to determine the benefits versus the costs of wetland development, thereby increasing the likelihood of achieving short-term development gains at the expense of a long-term decline in ecosystem service benefits. Worse still, it is the rural poor who are more likely to be burdened with the long-term consequences of wetland degradation.

In spite of developing one of the most advanced wetland policy environments in the world, the lack of financial and logistical support from central government for wetland management strategies at the local level has meant that the sustainability of wetlands in many areas rests on the development of community-based institutional arrangements. This chapter now goes on to examine the way in which wetland policy has evolved in Uganda, the lessons learned and, in particular, how community-based approaches have emerged as an important means of balancing the ESS of wetlands. The chapter draws on the example of the recent COBWEB, funded by the UNDP-GEF (United Nations Development Program – Global Environment Facility) and managed by the International Union for the Conservation of Nature (IUCN), to illustrate the ways in which communities can play a central role in developing their own wetland management strategies. Finally, it will identify specific lessons from this project that can feed into wider debates on decentralized, community-based wetland management approaches in the rest of Africa.

The evolution of wetlands policy in Uganda

Prior to the 1970s, seasonal wetlands and the margins of permanent wetlands in Uganda were traditionally used for grazing cattle, growing crops, domestic water supply and as hunting and fishing areas. Indeed, throughout the 1960s and 1970s a national policy that sought to double agricultural production and encourage agricultural expansion effectively encouraged the draining and conversion of wetlands. In Kigezi, for example, the government leased out large sections of the Kiruruma wetlands and provided subsidies to large-scale farmers to drain the wetlands and convert them into dairy farms (WID/IUCN, 2005). In Kampala, most wetlands were earmarked for industrial development and expansion of the capital city (Emerton, 2005).

During the 1980s, however, pressure on wetlands mounted in both rural and urban areas and many began to exhibit signs of degradation as a result of the various factors outlined above. The population trend soon after the 1970s led to increased pressure on the wetland resources to a level that was threatening the ecological, social and economic sustainability of Uganda's wetlands. The first important milestone and turning point was in 1986 when the National Resistance Movement government put a stop to the rampant and wanton conversion that characterized wetland management in the 1970s. The purpose of the ban was to await the development and adoption of a policy that would provide the basis for environmentally sound management and rational utilization of the wetlands. The ban on wetland drainage and the establishment of the Ministry of Environmental Protection the same year marked the beginning of the new era in wetland management in Uganda. This was followed by the ratification of the Ramsar convention in 1988 which further proved the government's commitment to promote sustainable management of wetlands.

With technical assistance from the IUCN, the National Wetlands Programme was created in 1989 specifically to develop a National Wetlands Policy that would guide the wise use of wetland resources in the country.

Soon, however, it was realized that a much wider approach focusing on public measures was required to achieve the goal of sustaining Uganda's wetlands. *The National Wetlands Policy* was developed in several steps, the first being a series of sensitization and awareness-raising campaigns, alongside consultations, to enlighten the public on the values and functions of wetlands alongside the need for their conservation and sustainable use.

Following a final round of public consultation and the approval of the policy by cabinet in 1994, the National Wetlands Policy was officially launched in 1995 with the overall aim to 'promote the conservation of Uganda's wetlands in order to sustain their ecological and socio-economic functions for the present and future wellbeing of the people' (MNR, 1995: 3). The policy then formed the basis for integrating wetland issues into key government legal frameworks, including the national constitution in 1995, the environment statute 1995, the local government act 1997 and the land act 1998. The constitution of the Republic of Uganda (ROU), chapter XIII states that 'the state shall protect important natural resources, including land, water, *wetlands*, oil, minerals, fauna and flora on behalf of the people of Uganda' (ROU, 1995: 16; emphasis added). Article 273 (2)(b) goes on to state that 'the government or local government as determined by parliament by law, shall hold in trust for the people and protect natural lakes, rivers, *wetlands*, forest reserves, game reserves, national parks and any land to be reserved for ecological and touristic purposes for the common good of all citizens' (ROU, 1995: 112; emphasis added).

In 1998, the WID was created as an institution for coordinating the implementation of the National Wetlands Programme. Initially situated within the Ministry of Water, Lands and Environment, the WID was later upgraded in 2006 to be the government WMD. One notable achievement of the WID was the development of the 'Kampala matrix' for wetland classification in 2000. This framework (Table 7.2) was developed to help classify wetlands by their relative importance and by their conservation status and threats. The classification helped prioritize, define and plan appropriate wetland conservation and management interventions, e.g. whether to restore a wetland, encourage wise use or simply monitor its use.

Table 7.2 The Kampala matrix

		Status		
		Threatened	*Not threatened*	*Destroyed*
Importance	*Vital*	Restore	Monitor strictly	Restore
	Valuable	Ensure wise use	Monitor	Restore?
	Dispensable	Encourage wise use	Monitor?	Forget for the time being

Source: adapted from MWLE/WMD, 2001

The policy provided a good foundation for the management of wetlands and succeeded in devolving their management to local government and the community level. The WID was linked to the local governments through regional technical support units that provided technical backstopping to the districts in implementing wetland management. However, the main challenge to wetland management still remains the limited capacity for management at the local level.

In order to support policy implementation, a number of guidelines were developed to guide wetland users in the type and intensity of activities that could be carried out in wetlands if their ecological functions are to be sustained. For example, when it was noted that the existing agricultural management practices in reclaimed wetlands and along wetland edges were not sustainable, the WMD in 2005 prepared guidelines to educate wetland cultivators to better and sustainably manage the wetlands under cultivation. Trees were being cut down indiscriminately in the wetlands to eradicate birds considered as pests, and the drainage and removal of vegetation was reducing the residence time of water in the wetlands. Wetland sedimentation was resulting in flooding and pollution of water sources (WID, 2005). Consequently, the WID urged caution to ensure that the hydrological and ecological support systems of the wetlands were not damaged or endangered by these agricultural activities. The strategy was to utilize the agricultural potential of the wetlands while at the same time maintaining their ecological integrity, the rationale being that poor wetland management can do tremendous damage to the economy, ecology and peoples' livelihoods.

The guidelines, for example, require that before conversion of the wetlands into cropland, users should seek advice from relevant technical authorities, and that developments covering more than 0.25 ha are not permitted unless an Environmental Impact Assessment (EIA) has been conducted and approved. Total drainage and conversion of wetlands is prohibited, and the guidelines require that areas of natural wetland must be retained between fields to sustain ecosystem functions and biodiversity. The guidelines also include specific regulations on how to manage water levels, crops, agro-chemicals and soils under cultivation. Critically they also recognize that people must be able to feed themselves now and in the future and should, therefore, be guided to use wetlands in a sustainable manner.

In 2001, the Wetland Sector Strategic Plan (WSSP) 2001–2010 was launched outlining the actions for implementation of the policy. The plan highlighted the importance of resource assessment, management planning and EIA capacity building, and was implemented at selected wetlands of 'vital' importance (see Table 7.3) where routine monitoring had been undertaken since 1997. The wetlands policy also drew on the various principles enshrined in successive Ramsar declarations that recognize the need for management to take into consideration the use of wetlands by people. It is for this reason that policy implementation guidelines for wetland edge gardening and paddy rice growing were developed and adopted. Thus the policy provides for

Table 7.3 Key events in the evolution of wetland management policy in Uganda

Year(s)	Key event
1970s	Wetland reclamation and drainage encouraged
1986	Large-scale wetland drainage banned
1988	Uganda ratifies the Ramsar Convention on wetlands
1989	National Wetlands Programme commenced
1994	National Wetlands Programme adopted
1995	Wetlands inscribed in national constitution
1995	Wetlands incorporated in the National Environment Statute (1995)
1997	Wetlands included in the Local Government Act (1997), decentralizing their management to local governments
1998	Wetlands Inspection Division (WID) born
1998	Wetlands included in the Land Act (1998)
2000	The 'Kampala matrix' for wetland classification developed
2000	Wetlands, Riverbanks and Lakeshore regulations (2000)
2001	Wetlands Sector Strategic Plan 2001–1010 launched
2005	Guidelines for wetland edge gardening and paddy rice cultivation adopted
2006	WID upgraded into a Wetlands Management Department (WMD)
2009–2011	National Wetlands Bill drafted, awaiting parliamentary approval
2011	Step-by-step guidelines for developing ordinances and byelaws for wetland management in Uganda prepared
2011	Strategy for effective wetland boundary demarcation drafted

Source: adapted from WID/IUCN, 2005

community-based natural resource management (CBNRM) since it requires negotiations and agreement with people on key activities to be undertaken in wetlands, and management arrangements that include people and ensure sustainability. The policy specifically has a key principle that requires that wetland conservation should be achieved through a cooperative approach involving all stakeholders including local communities.

Based on past experience and new emerging issues, Uganda has now drafted a National Wetlands Bill which awaits discussion and approval by parliament. In 2011, step-by-step guidelines for developing wetland ordinances, byelaws and a strategy for effective wetland boundary demarcation were prepared to respond to the emerging needs of decentralized wetland management.

Community management of wetlands

An increasing body of literature over the last 20 years has drawn attention to the significant contribution that community-based approaches can make towards sustainable natural resource management (NRM) (see Chapter 1; Fisher *et al.*, 2005). Concurrently there has also been increasing interest in the role of decentralized local institutional arrangements, i.e. those institutional structures to which the management of natural resources can be devolved. These can range from traditional community-based organizations (CBOs) (see Rasmussen and Meinzen-Dick, 1995; Leach *et al.*, 1999; Agrawal, 2001) to formally elected local government structures (Crook and Manor, 1998; Ribot, 2011). In both cases, the development literature has emphasized the potential of such local institutions to empower community participation and transfer decisions about resource use closer to those who use them. As highlighted by Ribot (2011), however, there is a significant emerging debate that questions the relationship between development interventions and decentralized institutions at the local level. Of particular concern has been the way in which external development agents, such as non-governmental organizations (NGOs) and central government, have sought to build parallel institutions that bypass existing locally respected authorities. According to Ribot *et al.* (2008), this has had the effect of undermining local representation by depriving legitimately elected governance structures of power, which instead is transferred to newly created institutions. Furthermore, this may create a situation whereby different institutions are in conflict as they compete with each other for limited material resources and political legitimacy. Despite recognition that many local authorities have a history of being non-responsive to local people, Manin *et al.* (1999) maintain that any external development intervention should, nonetheless, aim to empower local governments alongside grassroots community structures.

This is the political arena in which the decentralized, community-based management of wetland resources occurs, in Uganda and throughout Africa. In Uganda the inclusion of local people has been provided for in the National Wetlands Policy where one of the three key principles of the policy states that 'wetlands conservation can only be achieved through a coordinated and cooperative approach involving all the concerned people and organizations in the country, including the local communities' (MNR, 1995: 4). The approach is recognized as the most appropriate way forward for sustainable NRM and more likely to be successful than the strict protectionism that characterized the formation of the protected areas and Integrated Conservation and Development Projects (ICDPs) (Chapter 1; Fisher *et al.*, 2005). Whilst many external agencies claim to implement and promote CBNRM, in practice, it is often undertaken in a tokenistic manner and, as highlighted above, fails to work within existing institutional structures or facilitate the devolution of management to communities. As such, the results that can be achieved for both livelihood improvement and biodiversity conservation are limited.

In acknowledging these issues the Government of Uganda, since 2004, has opted to disburse part of its Poverty Alleviation Fund (PAF) to all District Local Governments, to target and operationalize decentralized wetland management structures. The total amount of funding a district receives is, among other criteria, linked to the number of community-based wetland management plans that have been developed and are under implementation in that district. The community-based approach to managing wetlands has also been incorporated in the draft wetlands bill that is currently before parliament for discussion. One example of the implementation of this approach is the COBWEB project.

Case study: the COBWEB project

Extending wetland protected areas through COBWEB was a UNDP-GEF-funded project, whose overall objective was 'to establish and strengthen community based systems and regulations that promote the sustainable use of wetlands with important biodiversity'. The project was a four-year (2008–2012) collaborative endeavour between the Government of Uganda (represented by the WPD) and an NGO consortium consisting of the IUCN, Nature Uganda (NU) and the Uganda Wildlife Society (UWS). The UNDP delegated implementation responsibility to the IUCN. The rationale of the project was drawn from the observation that despite Uganda's concerted efforts towards better management of wetlands, many remained vulnerable to degradation. The problem was exacerbated by the fact that wetlands remain under-represented in the National Protected Area (PA) Network; protected area coverage in Uganda has been heavily skewed to landscapes dominated by forest and savannah areas, leaving out the country's freshwater bodies and associated wetland ecosystems.

The project's approach drew from the IUCN 2008 categorization of protected area management and governance types that recognize Community Conserved Areas (CCAs) – run by traditional communities and local people – that are recognized and designated by government (Dudley, 2008). The aim of the project, therefore, was to develop, pilot and promote the adoption of a suitable wetland CCA management approach in two wetland systems adjacent to two terrestrial protected area networks: the Lake Mburo-Nakivale wetland system adjacent to the Lake Mburo PA in south-western Uganda, and the Lake Bisina-Opeta system adjacent to the Pian-Upe PA in north-eastern Uganda (Figure 7.1). This was to be achieved through three main objectives:

- Raising awareness of the links between wetland biodiversity and livelihoods.

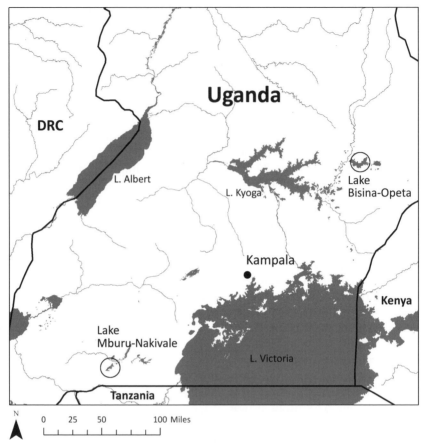

Figure 7.1 Location of the Lake Mburo-Nakivale and Lake Bisina-Opeta wetland systems

- Demonstrating and implementing wise-use practices.
- Integrating community-based conservation models into policy and planning.

The first focused on biodiversity conservation at each site and involved raising awareness of the value of the wetland biodiversity to livelihoods as a prerequisite to promoting sustainable use and biodiversity conservation. Awareness raising activities targeted a range of stakeholders including district level and local leadership, with the aim of ensuring that wetlands and biodiversity concerns were integrated into local-level planning processes as a means of entrenching their formal recognition and consideration for budgeting and management. Seven local governments have been supported to integrate six wetland CCA management plans into six subcounty development plans. CCAs that included specific plans for managing biodiversity as well as enhancing livelihoods were established at Magoro, Kapir, Mukura, Lake Nakivale, Kacheera I and Kacheera II. Included in these management plans were guidelines for local government staff on the development of wetland management byelaws.

COBWEB's second objective was to pilot and demonstrate wise-use activities in a manner that limits the loss of biodiversity values. Detailed biodiversity and socio-economic values of wetlands were recorded and mapped, building on community and Geographical Information System (GIS) maps that were used to monitor changes in both biodiversity and livelihoods. Ecological, socio-economic and Knowledge, Attitudes and Practices (KAP) baseline surveys were conducted, and their reports published and disseminated to inform strategic planning processes and promote wetland wise use (Nature Uganda, 2009; Amanigaruhanga and Lyango, 2010). It is against these baseline conditions that changes to biodiversity were measured to ensure that wise-use activities were implemented without a loss of the biodiversity function of the wetlands at the project sites. Annually, data were collected to support assessment of the sustainability of wetland use, with respect to biodiversity values and functions. Based on the project's action research, the best practices for sustainable wetland use were then developed, tested and promoted. Subsequently, a range of various wise-use activities are being demonstrated and promoted for adoption by communities at the CCA sites. These include wetland catchment management activities, such as tree planting to reduce lake sedimentation and the associated loss of aquatic biodiversity, but also sustainable fishing, wetland boundary and lake buffer zone demarcation, upland rice cultivation, sustainable fishing and ecotourism.

The third objective was to promote the wetland CCA approach itself, targeting policy makers and practitioners. This involved documenting and disseminating lessons learned and best practices from the project, and promoting their uptake, integration into policy and ultimately the proliferation of the CCA model. In doing so, one outcome of the project has been the development of a strategy for wetland demarcation for the whole country, which was undertaken by the Ministry of Water and Environment in 2010, and which has created an enabling framework under which CCAs can be promoted and implemented by the government. In order to mainstream wetland biodiversity concerns into the planning process and protected area networks, and base these on lessons learned, the project supported joint technical and policy advocacy workshops to promote recognition of the wetlands CCA approach. As a result, the draft wetlands bill and wildlife policy documents provide for recognition and adoption of the CCA model.

Mobilizing the community to manage wetlands

One key challenge for COBWEB in terms of promoting CBNRM lay in effectively disseminating the idea that communities should manage wetland resources for themselves from the very beginning of the process. One issue here is the land tenure policy; wetlands in Uganda are held in public trust by local government rather than being owned by individuals, hence it has been argued that this facilitates a 'tragedy of the commons' scenario whereby people compete for wetland use but do not take up the responsibility of management (Amanigaruhanga and Lyango, 2010). Traditionally, communal local management systems existed and appear to have facilitated the sustainable use of wetlands for water supply, grazing cattle, crop cultivation, hunting and fishing areas. As pressure on wetlands grew during the 1980s these traditional systems became less effective and, as highlighted previously, many wetlands started to exhibit signs of degradation. Furthermore, the local governments who own the wetlands in trust on behalf of the public, and who are mandated to provide technical guidance to communities on sustainable wetland use practices, were ill-equipped to take on the management role.

The first step towards mobilizing communities to participate in CBNRM was to understand the extent to which wetland resources contributed to peoples' livelihoods. Drawing on the Sustainable Livelihoods Framework (SLF) (see Chapter 1; Scoones, 1998), this involved identifying and assessing the enabling or disabling social and economic factors that determine how different communities (and individuals within communities) use wetlands, and also the range of livelihood options available to them. It involved understanding the various institutional structures, both formal and informal, that influence the use of wetland resources. Through understanding the social, political and economic context, the project was able to identify barriers and opportunities that could be used as entry points to facilitate a process to negotiate and plan for sustainable wetland management. Critically, these assessments were also sensitive to catchment and ecosystem level issues, and hence considered upstream and downstream interactions as well as land tenure arrangements, poverty levels, market trends and recurring shocks and pressures.

Because they were carried out in a participatory manner, these assessments also helped raise awareness and discussion among communities of the intimate links between their livelihoods and wetland resources. In particular, care was taken to ensure that each community appreciated and were sensitive to the livelihood needs of different people and how this affected the ways in which wetland resources were valued. Conversely, communities were given the opportunity to discuss their commonality in terms of wetland dependency, and how each individual's use of wetland resources potentially impacted on the livelihoods of others. One key theme emerging from all the participatory assessments was a consensus among stakeholders that livelihood strategies had been affected by the deterioration of the wetlands. Specific issues included a decline in fish catches due to the destruction of breeding sites by encroaching farmers, receding lake waters attributed to siltation from upland farms and an increase in the agricultural use of wetlands due to a decline in upland soil fertility and soil erosion. Due to its proximity to one of the largest refugee camps in the country, and the water abstraction associated with this, the waters of Lake Nakivale, in particular, have receded so much during the last decade that drinking water intake points have been moved closer to the lake on three occasions. Overall, these various outcomes of the participatory assessment successfully highlighted the links between wetland resources and peoples' livelihoods both at present, and in the future, and as such have empowered further community engagement in the management and restoration of their wetlands.

Identifying the entry points

Once the communities identified the need to restore wetland resources and recognized that each individual had a role to play in wetland management, the next step was to agree on the roles and responsibilities for this to happen. It was noted, however, that the incentive for the communities to participate had to be direct and linked specifically to immediate livelihood needs. Subsequent participatory activities, therefore, encouraged communities to develop long-term strategies (ten years) outlining how they could use wetland resources to enhance their wetland-dependent livelihood options. Communities were organized according to the common user groups, such as farmers and fishermen, and within these micro-planning units a long-term plan was discussed and developed. This included a process to collectively analyse and identify the barriers and incentives that influence the strategies developed. One noteworthy observation from this process was the eagerness of communities to engage in the future management of wetland resources, even to safeguard the less-tangible ESS, such as micro-climate regulation. Once the relationship between wetland resources and livelihood options became clear, a careful evaluation of the situation was undertaken and various groups were given the opportunity to negotiate between themselves and identify compromises that would benefit the entire community. For example, while it had been known that cultivation so close to the edge of wetlands was silting the lakes and destroying fish breeding sites, it had been difficult to convince people to move cultivation out of the wetlands since that was their main livelihood source. However, once the analysis and discussion showed that their livelihood would be compromised further if the practice continued, they agreed to the phased development of a buffer zone between their farmlands and the lake shore. Subsequently, the boundaries of the 200 m-wide buffer zone have been planted with stabilizing grevellia tree species and fruit trees that are useful for the landowners. What is significant here is that this has been achieved through a combination of community-based discussion of wetland resource dynamics and existing policy; the 200 m buffer zone is provided for in the law within the national lake shore management regulations developed by the National Environment Management Authority (NEMA) in 2000, which can be enforced by central and local government (MWLE, 2000).

However, it was also realized that these buffer zones could prove inadequate if soil erosion from the catchment was not addressed, and hence a different entry point had to be used to engage the landowners in the management of catchments. Landowners in the catchment are mainly agro-pastoralists who have experienced severe soil erosion in recent years as a consequence of overgrazing and crop production techniques that did not integrate soil and water conservation measures on the steep hill slopes. Discussions with landowners in the catchments revealed that very large gullies had formed due to successive years of rapid runoff. In spite of the presence of agricultural extension workers in the districts, landowners appeared powerless to deal with this problem; though for many the main barrier was simply a lack of knowledge about how to deal with them.

Consequently, the project arranged a field trip for some landowners that enabled them to visit a site that had similar characteristics to theirs but where soil erosion had been managed by establishing soil and water conservation structures across the hilly landscape of Kabale District. Based on the field trip, a road map on how to address the catchment soil erosion problems was drawn-up in consultation with landowners. Key to this was the agreement to set aside a day each week to work together and establish contour trenches that cut across the landscape. As a result new institutional arrangements for soil and water conservation were created. Following requests for assistance to the local district office, an extension agent was sent to provide technical support to farmers and to train a number of community facilitators on how to establish contour trenches. The objective of these trenches was to check and reduce the rates of surface runoff and soil erosion from the hillsides. The trenches prevented erosion by trapping runoff, allowing it to seep through the soils and also trapping soil that would have been washed further downslope. The trenches have been quite effective and, for this reason, have been adopted by three other farmers' groups in neighbouring villages. Although there is a need to follow-up these activities and assess their overall effectiveness, initial reports from farmers suggest that trapping and allowing runoff to seep into their fields has improved crop production, particularly banana productivity. It is important to note that although the entry point was to address the current erosion problem and increase upland productivity, both the upstream and downstream farmers appreciate the contribution of the soil and water conservation measures to the recovery of the lake and wetland system, in terms of access to cleaner water and improved fish stocks.

Once the various resource-use plans were drawn up, an interactive process was also facilitated where all the resource users came together and merged these plans. Based on the plans, community-level representatives from each user group were selected to create an oversight structure for the entire wetland ecosystem. In the structure each user group agreed on areas of protection and management that contribute to the health of the overall ecosystem. Furthermore, a monitoring and reporting system was drawn up and this included participatory checks among, and by, the various user groups, to assess each other's progress in implementation of their plans, holding each other accountable for success and failures.

Over the two-year period since the inception of COBWEB's participatory management plans, the Nakivale-Mburo Ramsar wetland site has registered some impressive results; in particular, satellite images from on-going PhD research indicate recovery and regeneration of wetlands at this site. Though the information is still anecdotal, communities are reporting better fish catches which implies recovery of the breeding sites. There has also been a marked recovery of the lake and because of this the previously abandoned water intake points have been submerged. Communities have also indicated that wildlife such as the crested crane and hippos have returned to the area even though the return of the hippo is considered a problem as they destroy the communities' boats. The biggest challenge, however, concerns the sustainability of the systems in place. It is likely that population growth will continue to exert more pressure on wetlands as people seek land for settlement, agriculture and industrial expansion. In recognizing this, the project's sustainability plan deliberately seeks to integrate the community-based management structures it has established into the more robust formal local government structures and seek their formal recognition. This is regarded as an important step towards legitimizing sustainable livelihood-related conservation activities such as ecotourism and sustainable fishing, whose tangible benefits will motivate further sustainable management practices.

Integration of the plan into policy

Anchoring community-based wetland management plans and management structures in local government institutions ensures that wetland sustainability is enshrined within a legal framework. The policy environment in Uganda provides a very good enabling environment for CBNRM, and indeed one strength of the project has been the participation of local government authorities from the very beginning. The IUCN and partners played a central role by developing the original objectives and identifying opportunities where communities could work with local government authorities in shaping policy. The timing of the latter was specifically chosen to coincide with the budgeting and planning periods for the districts so that district and national plans recognized the community initiatives and thus planned and provided resources to support them. In addition to these policy-influencing platforms, the management structures developed within communities were registered as CBOs to enable their formal recognition within government policy. For example, the institutional arrangements for managing wetlands, such as observing the buffer zone and protecting fish breeding grounds, were turned into byelaws that were agreed and recognized at the district level.

The project also convened annual policy meetings at the national level where outcomes of the project were presented and discussed. As a result of this process, project outcomes have been used to influence the proposed wetland bill that recommends the creation of community conservation areas following the COBWEB approach. Furthermore, the wildlife policy, which is under review, is considering the inclusion of support to the management of wetlands adjacent to PAs. This is to address issues such as the hippos which are considered as problem animals. The idea is to turn the challenges into opportunities by training community rangers who know how to handle the wildlife, as well as tapping into the tourism market of the park. This was, however, based on guidance that tourism development should not be at the expense of critical livelihood support systems; it had to be over and beyond the projected livelihood support emanating from the wetlands.

Overall, the important result to note is that whereas there were various entry points to engage the different categories of people based on their livelihood needs, it was the wetlands themselves that bound everyone together and that constituted the common framework for planning natural resource restoration. The early lessons have encouraged the community and the local government as well to scale up and build on the results of the project. At the Lake Nakivale site, for example, three new farmers' groups have emerged and are replicating the soil and water conservation measures in other villages within the wetland catchment area, and four other village councils are demarcating and regulating the use of wetland buffer zones. To manage the wetlands and the buffer zones, each of the six CCAs at project sites have management committees composed of representatives of the different wetland resource users (fishermen, cattle grazers, etc.). The buffer zones remain primarily under natural vegetation, cultivation is prohibited to reduce the risk of siltation and only controlled grazing is allowed. At the Kacheera CCA, livestock watering within open water is discouraged to maintain good-quality water for domestic use and livestock watering troughs have been built outside the lake shores to address this.

Conclusions and lessons learned

Though the final evaluation is yet to be conducted for the COBWEB project, there are already some key generic lessons learnt in terms of interacting with wetland users, implementing policies and effectiveness of CBNRM interventions.

Identifying appropriate institutional entry points for wetland governance

It is important to recognize, adopt and engage locally respected resource governance structures that promote local responsibility, belonging and representation. Ideally, the institution or individual selected should have the capacity to enhance free expression of views, a sense of belonging and representation of all community members. These structures should be linked to democratically elected formal governance structures that are mandated to manage wetland resources, but they should also incorporate mechanisms that render them accountable to both community members and higher governance levels. The choice of the management structure linking communities to government will invariably differ from site to site, since some areas are already predisposed to strong customary governance structures, whilst others are dominated by statutory ones. Similarly it is important to recognize the heterogeneity of communities in terms of human and social capital. At the Kapir CCA, for example, the 'champion' was a Beach Management Unit (BMU) Committee, whereas at Lake Nakivale it was two individuals – the District Wetlands Officer and an elderly farmer. At the Kacheera I CCA, meanwhile, it was a democratically elected woman councillor from the local area who was most influential in wetland management planning.

Participation with local communities

Acknowledging and respecting local communities and their rich knowledge of natural resource use, livelihoods and management issues are central to CBNRM. The role of the facilitators should simply be complementary and advisory. Wetland planning should involve participatory processes that enable discussions among all types of wetland users, and be representative of different stakeholder groups to enable an understanding of the links between resource management and livelihoods, the feasibility and sustainability of proposed changes to wetland management, and the existing opportunities and gaps in knowledge. Participation should also identify and respect communities' customary norms, rules and regulations. The COBWEB project was successful in adopting some positive aspects of communities' customary rules and regulations, and was able to transform these into byelaws that supported management of the wetland CCAs. The participatory nature of COBWEB also facilitated a unique partnership between government, NGOs and communities, which succeeded in re-orienting a top-down institutional focus away from the need to evict people from wetlands, to one where local communities were seen as having the potential themselves to solve wetland management problems.

It is also important to facilitate self-evaluation among the communities to assess and address their own capacity needs. The COBWEB project, for example, facilitated learning exchange visits to Kabale District for community members to learn hands-on skills in soil and water conservation; and to Bigodi wetlands, adjacent to the Kibale National Park for ecotourism. These participatory extension activities empowered them with necessary confidence and skills to spearhead new management interventions in their own communities.

Implement and inform existing policies

The COBWEB project sought to implement and entrench its various interventions through existing local and national policy, legal and planning frameworks. The first stage was to evaluate the policy, legal and planning frameworks that exist, discuss them with the communities and take advantage of them to implement the CBNRM project. In this sense, the COBWEB project sought to align itself with the National Wetlands Policy and the Wetlands, Riverbanks and Lakeshore regulations (2000), as well as decentralization policy and the local planning cycle to ensure that CCA management plans were integrated into local development plans and implemented. This ensured institutional sustainability and a continuity in resource allocation to the interventions through existing planning and budgeting processes. For example, Katakwi and Ngora District Local Governments and communities contributed land to build visitor biodiversity information centres, funds and materials for the improvement of access roads, and engaged in the construction of canoes to promote ecotourism. The fishing community at the Kacheera I CCA contributed the equivalent of US$1,041 to build a canoe for lake patrols to support sustainable fishing. This approach also helped translate lessons 'from policy to practice, and from practice to policy'; the draft National Wetlands Bill and draft National Wildlife Policy have, for example, adopted the wetland CCA approach.

The COBWEB policy formulation approach in which communities were given the opportunity to engage with their policy makers also enhanced the function and relevance of the district administration to the people. It also ensured that service delivery (e.g. technical environment and agriculture extension support from the local government) was guided to those areas that were most in need and had positive impacts. The local policy, legal and planning framework further provided the confidence and rules that encouraged people to participate in active management activities and also provided a basis to acquire support. These early lessons can indeed guide replication of these community-based wetland management approaches elsewhere. The area does not need to be a Ramsar site as in the Nakivale-Mburo case, and the facilitation role can equally be played by organizations other than the IUCN, by building upon such lessons learned.

Innovations for understanding wetland–livelihood linkages

For the wetland–livelihood linkage approach to be effective, the change has to start with the people. In many cases they need to be supported, often through practical demonstrations, to identify the direct links between sustainable wetland management and their livelihoods. Quite often, this link is under-estimated even by some members of local communities themselves. However, participatory planning, in which knowledge and practice is shared and discussed either within or between communities, has the potential to spur local sustainable development that balances developmental and environmental outcomes.

In a break with past wetland management policies in Uganda, COBWEB's primary focus was on the direct links between wetlands and livelihoods, and it is argued that through working directly with communities to identify and demonstrate these linkages, the project was successful in empowering a sense of ownership and collaborative management responsibility for wetlands. The COBWEB project also demonstrated that a range of entry points exist for intervention in the management of the wetland–livelihood nexus. At a fundamental (although often over-looked) level, baseline assessments (ecological and livelihood) can reveal entry points relating to challenges or opportunities that exist. Some may be very simple and require the removal of barriers to action, whilst others may be difficult and require synergies with other institutions. For example, the landowners at COBWEB sites just needed exposure to the practice of soil erosion control to address their problem, while baseline assessment revealed that fisher folk required support in the development of their market chains (Amanigaruhanga and Lyango, 2010).

The baseline survey process to understand the scope of wetland-based livelihood options, and how they are used and impacted on by different user groups, is of critical importance. Again, a key lesson learned from the project is that this should be undertaken in a participatory manner by the range of resource users themselves, simply because different people have different livelihood needs and aspirations. For the poor the most pressing issue on a daily basis is food security, while medium-level wealth groups may be more concerned with increasing their income. Resource management issues, activities and solutions will be different for each category, hence representational community engagement and participation is essential during the early stages of a project when surveying is required. This is also important throughout the later planning stages and in terms of monitoring the impacts of management interventions. For example, when communities established the first soil and water conservation trenches in the catchment, they were amazed at the impact one trench made in reducing the flow of the runoff, and because of these early achievements and their monitoring, more people as well as the local government were encouraged to participate further. Here, the evidence base, i.e. the demonstration of tangible benefits, played a significant role in maintaining the momentum and commitment of communities to the management interventions developed. At the Lake Nakivale CCA site the community monitors fish stocks for sustainable fishing, while at the Magoro CCA key bird species are monitored for ecotourism development.

However, it is important to recognize that the CCA model may not be effective for demonstration of the wetland–livelihood approach in extensive wetland systems. Managing huge wetland systems is quite complex on account of the diversity of people, communities and governance structures, and their needs and aspirations. Hence a single management structure, for example, may not be feasible and effective. It is recommended that smaller and manageable areas should be mapped out to be managed by communities while building upon the wetland–livelihood linkages. The smaller CCAs can then be consolidated later within a broader framework, the recognition can then be sought at higher governance levels.

In conclusion, the COBWEB project represents a significant step forward in the evolution of wetlands policy in Uganda. Above all, it has demonstrated that wetland management decision making can be re-orientated away from central government and successfully devolved to partnerships comprising local administrative institutions and communities. As highlighted throughout this chapter, this approach can have significant benefits not only in terms of sustaining the livelihood benefits from wetlands but also in enhancing the capacity of local institutional structures (formal and traditional) to respond directly to the needs of wetland users – in effect the intrinsic social and institutional dimensions of sustainable development. This will undoubtedly be of paramount importance in the years to come as the demands for wetland ESS increase in response to socio-economic and demographic change, but also as there are increasingly evident impacts of climate change. Having laid the foundation for sustainable wetland management, we argue, therefore, that the next stage in the development of COBWEB's approach should be a focus on the risks and vulnerabilities of wetland users and their livelihoods, leading to the development of mechanisms and institutions that build resilience and adaptive capacity at the community level.

References

Agrawal, A. (2001) 'Common property institutions and sustainable governance of resources', *World Development*, vol. 29, no. 10, pp. 1649–1672.

Amanigaruhanga, I. and Lyango, L. (eds) (2010) *A Socio-economic Baseline Survey of Communities Adjacent to Lake Bisina/Opeta and Lake Mburo/Nakivali Wetland Systems*, UNDP/GEF/ROU/IUCN, Kampala.

Crook, R. C. and Manor, J. (1998) *Democracy and Decentralization in Southeast Asia and West Africa: Participation, Accountability and Performance*, Cambridge University Press, Cambridge.

Dudley, N. (ed.) (2008) *Guidelines for Applying Protected Area Management Categories*, IUCN, Gland.

Emerton, L. (ed.) (2005) *Values and Rewards: Counting and Capturing Ecosystem Water Services for Sustainable Development*, IUCN Water, Nature and Economics Technical Paper no. 1., IUCN, Colombo.

Fisher, R. J., Magginnis, S. and Jackson, W. J. (2005) *Poverty and Conservation: Landscapes, People and Power*, IUCN, Gland.

Glass, S. (2007) 'Implementing Uganda's National Wetlands Policy: a case study of Kabale District', http://digitalcollections.sit.edu/cgi/viewcontent.cgi?article=1131&context=isp_collection (accessed 30th July 2012).

Howard, G. W., Bakema, R. and Wood, A. P. (2009) 'The multiple use of wetlands in Africa', in E. Maltby (ed.) *The Wetlands Handbook*, Blackwell, Oxford.

Leach, M., Mearns, R. and Scoones, I. (1999) 'Environmental entitlements: dynamics and institutions in community-based natural resource management', *World Development*, vol. 27, no. 2, pp. 225–247.

Manin, B., Przeworski, A. and Stokes, S. (1999) 'Elections and representation', in A. Przeworski, S. Stokes and B. Manin (eds) *Democracy, Accountability and Representation*, Cambridge University Press, Cambridge.

MFPED (2004) *Poverty Eradication Action plan (2004/5-2007/8)*, Ministry of Finance, Planning and Economic Development (MFPED), Kampala.

MNR (1995) *National Policy for the Conservation and Management of Wetland Resources*, Ministry of Natural Resources, Kampala.

MWLE (2000) *The National Environment (Wetlands, River Banks and Lake Shores Management) Regulations*, no. 3/2000, Ministry of Water, Lands and Environment (MWLE), Kampala.

MWLE (2009) *Water and Environment Sector Performance Report*, Ministry of Water, Lands and Environment (MWLE), Kampala.

MWLE/WID (2001) *Wetland Sector Strategic Plan 2001–2010*, Ministry of Water, Lands and Environment (MWLE), Kampala.

Nature Uganda (2009) *Ecological Baseline Surveys of: Lake Bisina-Opeta Wetlands System; Lake Mburo-Nakivali Wetlands System*, Nature Uganda, Kampala.

NEMA (2001) *State of the Environment Report for Uganda, 2000/2001*, National Environment Management Authority (NEMA), Kampala.

NEMA (2008) *State of the Environment Report for Uganda 2008*, National Environment Management Authority (NEMA), Kampala.

Rasmussen, L. N. and Meinzen-Dick, R. (1995) *Local Organizations for Natural Resource Management: Lessons from Theoretical and Empirical literature*, EPTD Discussion Paper no. 11, International Food Policy Research Institute, Washington, DC.

Ribot, J. C. (2011) 'Choice, recognition and democracy effects of decentralisation', Working paper 5, Swedish International Centre for Local Democracy, http://sdep. beckman.illinois.edu/files/ICLD_VisbyWorkingPaper_05.pdf (accessed 30th July 2012).

Ribot, J. C., Chhatre, A. and Lankina, T. V. (2008) 'Institutional choice and recognition in the formation and consolidation of local democracy', Representation, Equity and Environment Working Paper no. 35, World Resources Institute, Washington, DC.

ROU (1995) 'The Constitution of the Republic of Uganda', http://www. ugandaembassy.com/Constitution_of_Uganda.pdf (accessed 30th July 2012).

Scoones, I. (1998) *Sustainable Rural Livelihoods: A Framework for Analysis*, IDS Working Paper 72, IDS, Brighton.

UBOS (2012) 'Statistics Abstract 2012, Uganda Bureau of Statistics, Kampala', http://www.ubos.org (accessed 30th July 2012).

WID (2001a) 'Guidelines for smallholder paddy rice cultivation in seasonal wetlands', Wetland Booklet no. 3, Wetlands Inspection Division, Kampala.

WID (2001b) 'General guidelines for wetland management', Wetland Booklet no. 6, Wetlands Inspection Division, Kampala.

WID (2005) 'Guidelines for wetland edge cultivation', Wetland Booklet no. 11, Wetlands Inspection Division, Kampala.

WID/IUCN (2005) *From Conversion to Conservation: Fifteen Years of Managing Wetlands for People and Environment in Uganda*, WID and IUCN, Kampala and Nairobi.

WMD/NU (2008) *Implementing the Ramsar Convention in Uganda: A Guide to the Management of Ramsar Sites in Uganda*, WMD and Nature Uganda, Kampala.

WMD, UBOS, ILRI and WRI (2009) *Mapping a Better Future: How Spatial Analysis Can Benefit Wetlands and Reduce Poverty in Uganda*, World Resources Institute, Washington, DC and Kampala.

8 Managing a Ramsar site to support agriculture and fisheries

Lake Chilwa, Malawi

Daniel Jamu, Lisa-Maria Rebelo and Katherine A. Snyder

Summary

Lake Chilwa, a Ramsar wetland of international importance, is a major source of fish and agricultural produce. As such it is vital for the livelihoods of large numbers of people who live in the surrounding area and also for urban populations in Malawi. The wetland also plays a critical conservation role as it supports a population of 1.5 million residents and non-resident waterfowl, and is both a breeding ground and resting and feeding station along an important bird flyway in Southern Africa. The productivity of the agricultural systems in the wetland depends on adequate rainfall to annually recharge the lake and maintain the water balance in the wetland. When lake levels are high it is one of the most productive lakes in Africa. A number of different management initiatives have been developed to try to balance human and conservation requirements. The last was developed in 2001 but has not been implemented in a holistic manner due to delays in the decentralization of government activities to the district level and lack of funding. This review of the Lake Chilwa management experience suggests the need for more effective decentralized institutional arrangements that fully engage the local communities and stakeholders, and recognize the need for trade-offs in ecosystem services (ESS) in the wetland and the catchment.

Introduction

The Lake Chilwa wetland is a major source of water, fish and agricultural produce which contributes to the livelihoods of many of the 1.5 million people that reside in the Lake's basin and many millions more outside. The wetland also supports a large number of nationally and internationally important flora and fauna, including 153 species of resident and 30 species of migratory Palaearctic birds. It was designated as a Ramsar wetland site of international importance in 1996 (Njaya *et al.*, 2011). The Lake Chilwa wetland is unique in Malawi because the lake undergoes periodic drying cycles of 15–20 years. The productivity of the lake and the fortunes of the people in the wetland change in concert with changes in the lake's water level. The productivity of the agricultural and fish systems in the wetland depend in particular upon adequate rainfall and inflows to annually recharge the lake and maintain the water balance of the wetland. When lake levels are high it is one of the most productive lakes in Africa, with fish yields (159 kg/ha/yr) similar to those from extensively managed aquaculture systems (Kalk *et al.*, 1979; Balarin, 1985).

Population growth of 3 per cent per annum (NSO, 2008) and high levels of poverty are prompting agricultural expansion and clearance of natural vegetation in the catchment, putting Lake Chilwa's resources under increased pressure. Farmers are facing declining agricultural yields from decreased soil fertility and are expanding cultivation in the upper catchments of the basin, resulting in deforestation which in turn increases soil erosion, rapid runoff and flooding downstream (Jamu *et al.*, 2003; Njaya *et al.*, 2011). The erosion results in greater siltation in rivers, clogging downstream irrigation systems and potentially damaging fish spawning areas in the rivers and lake (Environmental Affairs Department, 2000; Jamu *et al.*, 2003). In addition, studies show that increasing frequency and intensity of flooding and drought in the basin (which are also dependent on seasonal and inter-decadal rainfall patterns) are affecting the basin's productivity and resources (Njaya *et al.*, 2011).

In response to these pressures and to fulfil the expectations of the Ramsar Convention, a management plan was developed in 2001 as a collaborative effort between the Danish International Development Agency (Danida) funded Lake Chilwa Wetland and Catchment Management Project and Malawi's Environmental Affairs Department (Environmental Affairs Department, 2001a). The plan proposed a comprehensive and multi-sectoral approach to managing the catchment and lake which focused on treating natural resources as economic goods, as well as devolving control and management of resources to the community wherever possible. Unfortunately, Danida withdrew from Malawi before the second phase of the project when the plan was to be implemented. Thus, much of the progress that had been made as a result of stakeholder and government capacity building came to a halt in 2001. While the three districts that surround Lake Chilwa developed sectoral plans for fisheries, forestry, land use, agriculture and water, and non-governmental institutions that have implemented a variety of agricultural and forestry projects in the basin, there has been a lack of coordination. In addition, there has not been enough attention given to improving livelihoods and thus reducing poverty through improved natural resource management (NRM) or the improvement of off-farm labour opportunities. There remains a need for greater integrated management of the wetland that would take into account the wider environmental and socio-economic conditions that affect the lake (van Zwieten and Njaya, 2003; Njaya *et al.*, 2011).

In this chapter, we provide an overview of the Lake Chilwa wetland with particular reference to the livelihood benefits that people derive from the wetland. Current pressures on wetland resources, existing institutional arrangements for managing wetland resources and the challenges faced by managers in balancing development and conservation needs in the wetland are discussed. Finally, recommendations are made in relation to how the wetland can be sustainably managed through the development and implementation of integrated and multi-sectoral management plans.

Lake Chilwa wetland

Site description

The Lake Chilwa basin (Figure 8.1) is located in a tectonic depression in south-eastern Malawi at 15°30'S and 35°30'E. Based on the catchment boundary derived from elevation data acquired during the Shuttle Radar Topography Mission (SRTM), the Lake Chilwa catchment covers an area of 8,784 km^2 of which 5,724 km^2 (65 per cent) lies in Malawi and 3,060 km^2 (35 per cent) in Mozambique (Rebelo *et al.*, 2011). The lake, with an area of approximately 2,284 km^2, has a radial drainage pattern with 14 streams converging from the surrounding watershed, and is a closed system with no outlet. The lake catchment is bounded to the west by the Chikala Hills, the Zomba and Malosa Mountains, the Shire Highlands and Chiradzulu Mountain. These mountains give rise to the Zumulu, Lingoni, Domasi, Naisi, Mulunguzi, Likangala, Thondwe, Namadzi and Mombezi Rivers. The catchment is bounded to the south-east by the Mulanje and Michesi Mountains that give rise to the Phalombe River and the Sombani River which forms the Mpoto Lagoon before flowing into the lake (Njaya *et al.*, 2011). The eastern mountains in Mozambique give rise to the Bungwe, Mnembo, Matchimaze, Namajete and Cocole Rivers and the Eastern Rift gives rise to the Mikoko and Naminga Rivers. Average annual inflow to the lake (comprising both surface flow and any groundwater inflow) has been estimated to be 1,472 mm^3; 60 per cent of the water in the wetland originates as direct rainfall onto the wetland, with 40 per cent originating as flow from the surrounding catchment (Rebelo *et al.*, 2011). This equates to the equivalent of an average annual contribution from the catchment of 244 mm which corresponds to 18 per cent of the average annual rainfall.

The Lake Chilwa ecosystem consists of a shallow, enclosed endorheic saline lake with an average depth of 1–2 m and a maximum depth of 5 m in the south-east. The lake and its adjoining marshes, swamps and grassland floodplains together are approximately 40 km wide and 60 km in length. A recent map of the wetland derived from satellite data shows the site to consist of three distinct components: open water (828 km^2), seasonal open water (303 km^2, which is flooded in the wet season but at lower lake levels during the dry season consists of mud and macrophytes), and seasonally inundated marsh/grassland floodplain (688 km^2) which forms a belt around the former two (Rebelo *et al.*, 2011). The hydrology of the lake is an important control on the ecology of the wetland, influencing not only the water chemistry and physical properties but also the soil characteristics and composition of

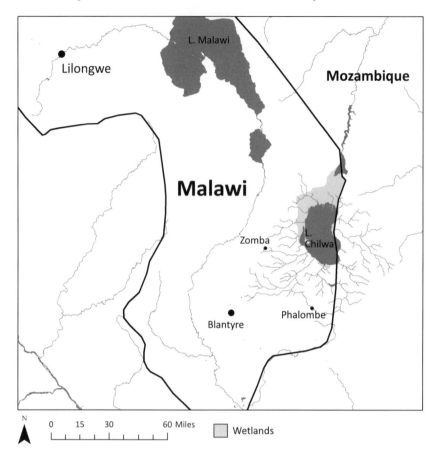

Figure 8.1 Map of the Lake Chilwa basin

Source: adapted from Njaya *et al.*, 2011

the vegetation (Howard-Williams and Walker, 1974). Lake Chilwa has experienced nine minor to moderately severe periods of declining water levels since 1900. The most recent episodes of complete drying occurred in 1967 and in 1995 as a result of drought (Kalk *et al.*, 1979; Njaya 2001).

In 1996, 2,284km² of the lake and surrounding floodplain were designated by Malawi as a Ramsar site under Criterion 6, i.e. for regularly supporting 1 per cent of the individuals in a population of one species or subspecies of waterbird (Rebelo *et al.*, 2011). (This designation does not include the area of wetland which lies in Mozambique, see Figure 8.1.) Among the important waterbirds are the glossy ibis (*Plegadis falcinellus*), Fulvous whistling duck (*Dendrocygna bicolor*), black crake (*Amaurornis flavirostris*), Allen's gallinule (*Porphyrio alleni*), lesser moorhen (*Gallinula angulata*) and grey-headed gull (*Larus cirrocephalus*).

Njaya *et al.* (2011) provide a comprehensive and up-to-date review of the status of the wetland's resources and the social context of the Lake Chilwa basin. Their review notes that the Lake Chilwa basin is one of the most densely populated areas in Malawi with approximately 164 people per square kilometre. Poverty is pervasive with 70–80 per cent of the population around Lake Chilwa living below the Malawi Government poverty line (US$0.41 per day) and 36–50 per cent of the Lake Chilwa population living in extreme poverty (US$0.25 per day) (Government of Malawi and IFPRI, 2002). According to a recent study the literacy levels in the basin are low with approximately 28 per cent of inhabitants not knowing how to read and write; 14 per cent without any formal education and only 16 per cent with secondary level education (Nagoli, 2010). As in the lake area, the population of the catchment is predominantly rural with 60 per cent dependent on agriculture for their livelihoods (Zimba and Kaunda, 1999).

ESS and resource use

The value of wetlands for people arises from the interaction of the ecological functions they perform with human society. In Africa, wetlands play a particularly vital role in directly supporting and sustaining livelihoods through the provision of a range of ESS that bring both physical and non-physical benefits to people. Frequently described as one of the most productive lakes in Africa, land, water, forestry and fisheries resources form the backbone of the Lake Chilwa wetland and wider catchment economy with over 85 per cent of households dependent on these resources for income and livelihoods (Nagoli, 2010). The right-hand panel of Figure 8.2 shows the main ecosystem components of the wetland (inner circle), the ESS provided by each of these (middle circle) and the ESS provided by the integrated system (outer circle). Fishing takes place in the area of permanent open water all year round while the floodplain is also used for fishing during the wet season. During the dry season, as flood levels recede, the floodplain is predominantly used for small-scale rice growing, grazing and the cultivation of vegetables. In addition, several large-scale irrigation schemes were established within the wetland in the 1970s growing high-yielding varieties of rice.

The high productivity of Lake Chilwa makes it critical to the economy of the entire basin, as well as making a significant contribution to the wider economy of Malawi. The economic benefits derived from the wetland are summarized in the inner circle of the left hand panel of Figure 8.2, with the total economic value estimated to be US$21 million per year (Schuijt and Brander, 2004). Of this it has been estimated that fish contribute US$18.7 million per year in net benefits, the grasslands – through grazing – provide US$638,000, the open water – through the jobs derived from canoe and boat transportation – US$402,557, the vegetation and clay – used for crafts and construction – US$14,000, while the agricultural

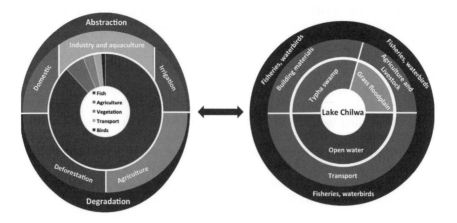

Figure 8.2 Major components of the wetland, economic benefits derived from the wetland and the threats to identified economic benefits

Note: left-hand panel shows the economic benefits derived from the wetland (inner circle) and the practices in the catchment (intermediate circle) which are seen as threats (outer circle) to the wetland; right-hand panel shows the major wetland components (inner circle) and the main livelihood activities which are practised in each of these (outer circles).

grounds (including crop-growing activities and small-scale organized rice schemes in the wetland areas) have been valued at US$1.2 million per year (Schuijt, 1999, 2005). In addition to water for irrigation, the lake is used as a domestic water supply. While income from tourism (e.g. birdwatching) is currently insignificant, the wetland also supports the livelihoods of over 400 bird hunters who trap 1.2 million birds per year valued at US$215,000 (Environmental Affairs Department, 2000).

Of particular note is the fact that the value of fish is ten times higher than rice, the second most important economic activity within the basin; per square kilometre fishing provides greater economic returns than either agricultural production or livestock grazing. Yielding as much as 25,000 tons of fish per year, the lake contributes between 16–43 per cent, with an average of 22 per cent, to the total fish catch of Malawi (van Zwieten and Njaya, 2003). However, the contribution of Lake Chilwa to the total fish catch of Malawi has declined over the past five years (Government of Malawi, 2008).

Pressures on ESS and resource use

The lake ecosystem is under increasing pressure as the human population grows with the expansion of agriculture (in both the catchment and the floodplain) and intensified fishing. The productivity of the agricultural systems in the wetland depends on adequate rainfall to maintain the annual recharge of the lake. Increasing water and energy demands in the catchment are increasing pressure on the lake and wetlands; two of the major current threats to Lake Chilwa are degradation in the catchment and reduction in lake levels reportedly due to abstraction for irrigation (Figure 8.2; Schuijt, 2005).

The degradation of land in the catchment has two main sources: deforestation and agriculture. The fisheries sector is a driver of deforestation, as fish smoking activities consume about 6,500 tons of firewood annually (Kabwazi and Wilson, 1998). In addition to fish smoking, illegal cutting of trees in forest reserves to supply firewood and charcoal to urban areas of southern Malawi has contributed to rapid deforestation in the catchment. Charcoal production is another major livelihood activity in the Lake Chilwa catchment (Chiwaula and Chaweza, 2011). If undertaken in an unsustainable manner this results in deforestation and associated land degradation, and in turn contributes to increased soil erosion and siltation of the lake. The most recently available national estimates for deforestation are 3.2 per cent per year (Environmental Affairs Department, 1999). Deforested and degraded areas within the catchment and cultivation without adequate soil and water conservation lead to high soil erosion rates (113 tons per hectare per year), resulting in declining agricultural productivity and increasing siltation of the rivers and wetland, as well as increased flooding throughout the catchment (Calder *et al.*, 1995; Jamu *et al.*, 2003). The intensity of flash floods has increased with deforestation in the catchment (Environmental Affairs Department, 1999) resulting in increased erosion. In addition, farmers complain of damage to irrigation infrastructure while road damage reduces farmers' access to input and output markets. Furthermore, the damage to water intakes and canals reduces the amount of water available for irrigation. While these impacts have not been economically quantified, an example of the scale of damage is evident from a flash flood and landslide in Phalombe in 1991 which resulted in the loss of 700–1,000 lives, and the destruction of 30,000 ha of cropland and ten bridges (Poschinger *et al.*, 1998; Cheyo, 1999).

Current land-use practices and consequent land degradation within the catchment are seen by fisheries managers as more of a threat to the resilience of the Lake Chilwa fishery and to peoples' livelihoods than over-fishing or the use of the lake's resources (Njaya *et al.*, 2011). Low water levels and dissolved oxygen, and high silt loads, alkalinities, pH and temperatures have been linked to mortalities in Lake Chilwa fisheries (Furse *et al.*, 1979; Macuiane *et al.*, 2011) and these stressors can be directly related to changes in watershed management and weather variability (Chimatiro and Vitsitsi, 1997; Jamu *et al.*, 2003).

Due to inadequate monitoring, reliable data are lacking but a widely perceived threat to the wetland is the abstraction of water for various uses within and outside the floodplain. Rice is the major crop grown in the floodplain irrigation schemes, while vegetables and maize are the major crops grown in the irrigation areas located outside the floodplain areas. There are three major irrigation schemes (Domasi, Likangala and Zumulu) and 12 small irrigation schemes in the floodplain area of the wetland with a total irrigable area of about 1,500 ha. The water abstraction by formal irrigation schemes per day in the 1990s was around 150,000 m^3 (Environmental Affairs Department, 2000).

Six new schemes ranging in size from 33 to 400 ha are planned in the basin, which will further increase the amount of water abstracted for irrigation. The water requirements for the new irrigated areas in the basin are estimated to range from 0.5 m^3/s for the small schemes (30–70 ha) to 1 m^3/s for the large (400 ha) schemes. Water abstraction by private estates, fish farming and informal schemes in the basin have not been quantified because most of these abstractions were initiated after the Water Resources Board, which was responsible for licensing water abstractions throughout Malawi, became dysfunctional in the 1990s.

Although based on very short time series, river flow data obtained by the Lake Chilwa Climate Change Adaptation Programme (established in 2010) show that the mean dry season discharge for most rivers in the catchment (i.e. 0.3–0.8 m^3/s) may be lower than the design water requirements of the new irrigation schemes (Water Department, 2010; LCBCCAP, 2011). Beusekom (2011) made similar observations on the inability of the Domasi River to satisfy dry season irrigation water demands of the Domasi irrigation scheme. The fact that irrigation schemes may have been designed inappropriately highlights the vital need for good data for the management of water (and other) resources. However, since Lake Chilwa is a closed basin with the vast majority of the flow entering the lake in the wet season, it is unlikely that dry season withdrawals will significantly affect lake water levels. Nonetheless, given the rapidly increasing withdrawals, it is clear that more research is required to confirm the exact nature of this threat to the wetland, particularly in dry years.

In addition to these formal irrigation schemes, there are also numerous small patches of land being irrigated for beans and horticultural crops by smallholders using treadle pumps, river diversions and watering cans (Nagoli, 2010). Because of the increased frequency of drought and dry spells within normal rainfall seasons, Malawi government policy promotes irrigation through the development of new irrigation schemes, the distribution of treadle pumps and the provision of subsidies for motorized pumps. In addition, the government's Greenbelt Initiative aims at reducing Malawi's dependency on rain-fed agriculture through the use of available water resources for irrigation in order to increase production, incomes and food security at both household and national levels for economic growth and development (Government of Malawi, 2010). The Phalombe Plain in the Lake Chilwa basin is one of the areas targeted by the Greenbelt Initiative. Through this initiative and other activities in line with the new policy, the number of large and small irrigation schemes is expected to increase, thereby putting additional pressure on water resources in the basin.

During periods of lake recession, fish production declines significantly, resulting in people turning to other livelihood activities. As these people look for other livelihoods, pressure on other basin resources, from birds to trees, increases to meet the need for cash income and food (Allison and Mvula, 2002). For example, in years when lake levels and fish production are low, bird catching increases by 300–500 per cent (Bhima, 2006).

Although most of the Lake Chilwa fish species are tolerant of a wide range of environmental factors and have the ability to adapt quickly to a changing environment, they are very dependent on the ecological condition of the wetlands. The marsh and floodplain surrounding the lake serve as critical breeding areas. However, burning of vegetation for hunting and to regenerate new grass for cattle grazing, as well as to improve access to the lake during the dry season, along with siltation and conversion of the floodplain to rice fields, all threaten the sustainability of the fishery (Njaya *et al.*, 2011, Rebelo *et al.*, 2011). Failure of the lake to seasonally inundate the extensive marsh and grassland floodplain greatly reduces the productivity of the lake and, in particular, the fisheries. In addition, the expansion of cultivation into the marshland may result in damage to fish breeding areas.

Closed basins such as Lake Chilwa are extremely sensitive to climate variations and changes in inflows and evaporation. As an emerging and future threat to Lake Chilwa, climate change projections for the region indicate a decrease in available water resources (IPCC, 2007) while some studies indicate that the surface air temperature in the Chilwa basin will increase by 2.6–4.7°C by 2075 (Chavula, 1999). Sarch and Allison (2000) note the importance of climate in driving the dynamics of fish stocks in African inland lakes such as Lake Chilwa.

Poverty and population pressures have significant impacts on the utilization of wetland resources and there is interdependence between different land-use practices and livelihoods across the basin. For example, fish smoking contributes to deforestation which increases the incidence of flooding and flash floods, resulting in soil erosion and reduced agricultural productivity. The increased demand for irrigation to reduce the population's vulnerability to climate change and increase agricultural production is leading to increased water abstraction and conversion of floodplain areas into agricultural zones. This in turn may result in changes to the wetland hydrological regime which can affect fish production by threatening the critical marsh and river spawning areas. There may be the creation of a vicious cycle of increasing pressure on wetland resources, declining productivity and increasing poverty. The situation is further exacerbated by conflicting policies and development plans (Nagoli, 2010).

Management challenges and trade-offs between resource use

While Lake Chilwa is an extremely valuable economic resource to Malawi and is critical to the livelihoods of the basin's population, it clearly faces great management challenges. The Ramsar 'wise-use' principle recognizes that wetlands, through their ecological and hydrological functions, provide invaluable services, products and benefits to human populations. As such, the challenge in managing a Ramsar site, which supports millions of livelihoods directly and indirectly, is to ensure that the wetland continues to provide the ESS that support these livelihoods for future generations, while at the same time conserving biological diversity and ecosystem health. The objective in managing the wetland and catchment should be to produce maximum net benefits for people whilst at the same time avoiding fundamental ecological threats and ensuring the long-term sustainability of the different ESS (Senaratna Sellamuttu *et al.*, 2008). The management of the wetland needs to take into account the trade-offs between activities taking place in the catchment and those in the wetland itself, and also within each area, to ensure the sustainability of the wetland. Analysis of the proportional contribution of the main economic benefits derived from the Lake Chilwa resources (the centre of Figure 8.2 – left-hand panel) indicates that fisheries and agriculture contribute to 93 per cent of the total net benefits. While these activities provide the major economic benefits, both are threatened by the activities in the catchment.

Although small-scale subsistence agriculture may only cause relatively small changes in other ESS, any agricultural activity within a wetland will alter its ecological character to some extent and there will thus be associated trade-offs. While the agricultural development of wetlands results in an increase in the provision of food in the short term, in the long term it often increases the input of pollutants, removes the natural filtering function of wetlands and reduces other ESS – both provisioning and non-provisioning (McCartney *et al.*, 2010). Hence, care is needed in agricultural management and incentives are required to ensure that negative impacts on other ESS are minimized. Within Lake Chilwa, households who have access to wetland gardens and grow a range of subsistence crops in the dry season reported that this farming enables them to produce enough food to last throughout the year. This results in an increasing demand for wetland gardens which causes conflicts of interest among the different wetland users and increases competition for water (Kambewa, 2005). In addition to the wetland vegetable cultivation, the increase in rice cultivation leads to a growing demand for irrigation water. At the same time, agricultural production within the wetland is vulnerable to changes in weather conditions and to the impacts from the various activities taking place in the surrounding catchment. Indeed, the sustainability of agricultural production on the floodplain is reliant upon the availability of water, with yields dropping dramatically during periods of low lake levels because residual moisture is depleted (Chavula, 1999).

Fishermen blame increased siltation on the upstream users who practise deforestation and poor land management (Mulwafu, 2000). The increasing siltation, along with localized burning and conversion to rice fields, are threatening the pools and marshes around the edges of the lakeshore, which are critical for fish breeding and for the survival of fish stocks when lake levels are low (Njaya *et al.*, 2011). Thus in order to maintain the productivity of the Lake Chilwa fishery there may be conflicts to resolve between livelihood activities within the catchment and those in the wetland.

The bird population of Chilwa also has a high economic value and over-exploitation is a major problem (Schuijt, 2005). This situation is worsened by the fact that the bird breeding season corresponds to the closed season for fishing on Lake Chilwa when many households experience food shortages. This confluence of factors is another challenge to address (Environmental Affairs Department, 1999).

Within the Lake Chilwa basin, there is clearly a need to manage the wetland for multiple ESS and to align these with livelihood strategies. Although in many cases this means a greater emphasis on sustainable use to protect key wetland functions, managing wetlands for livelihoods is not necessarily congruent with managing them solely to protect biodiversity (McCartney *et al.*, 2010). In the case of Lake Chilwa sustainable management will involve addressing conflicts and achieving trade-offs between livelihood requirements and conservation needs that require skilful and innovative forms of management to overcome. Conservation of ESS and functions are of particular importance under future climate change scenarios when the wetland may be a vital component of the adaptation strategies for the communities from Lake Chilwa and its catchment.

Institutional arrangements for management of Lake Chilwa

The lake and its natural resources are managed mainly through traditional and government systems (Njaya, 2007). In Malawi, traditional authorities, such as chiefs, serve as custodians of natural resources (Jamu *et al.*, 2011) and traditional beliefs and social norms shape the way people utilize these resources (Munthali, 1997; Lowore and Lowore, 1999). For example, it is customary to respect trees in graveyards, to preserve sacred groves or sanctuaries, and to leave certain tree species when opening a garden (Lowore and Lowore, 1999). In the Lake Chilwa fishery, certain areas are declared sanctuaries or closed areas where fishing is banned (Njaya, 2009). Use of traditional and customary practices to directly or indirectly control the exploitation of natural resources is becoming increasingly rare in various parts of Africa. In the formal government system, natural resources exploitation is controlled through a wide range of regulations (Lowore and Lowore, 1999; Jamu *et al.*, 2011; Njaya *et al.*, 2012). Government agencies such as Fisheries, Water, Wildlife and Forestry have put in place several sector-specific tools to help achieve sustainable resource management.

Lake Chilwa Management Plan

During the 1990s, the Danida-funded Lake Chilwa Wetland and Catchment Management Project worked with Malawi's Environmental Affairs Department and a wide variety of stakeholders to design an overall management plan for the Lake Chilwa basin. The project was formally launched in 1999 and included numerous components aimed at building the capacity of district staff, extension personnel and local communities in wetland management. It placed considerable emphasis on stakeholder engagement and promoted community-based natural resource management (CBNRM). Danida sponsored several studies to assess the state of the basin's natural resources as well as the status of people's livelihoods and health. These studies were compiled into a Lake Chilwa State of the Wetland Environment Report (Environmental Affairs Department, 2000). This report was then used as the basis for the design of the Lake Chilwa Wetland Management Plan (Environmental Affairs Department, 2001a).

The plan contained frameworks for environmental management, human development, soil and land management, forestry, fisheries, water, biodiversity and environmental education and awareness (Environmental Affairs Department, 2001a). The state of the wetland report identified specific environmental issues that were then addressed in the plan through 66 different strategies and 85 associated activities across multiple sectors. The plan was presented to the three relevant districts (Machinga, Zomba and Phalombe) within the basin for feedback and was modified based on their responses. It was intended that the management measures outlined in the plan would be adopted by each district and be incorporated into each district's Environmental Action Plans that are supposed to be created and revised every five years (Environmental Affairs Department, 2001b). Implementation of activities was envisaged to occur through area development committees (ADCs) and village development committees (VDCs) (Figure 8.3). These structures were developed as part of the Local Government Act (Chiweza, 2010). However, the second phase of the Danida project, which aimed to put the plan into action, did not go forward.

Decentralisation of natural resource management in law and practice

Prior to, and alongside, the initiatives of the Danida project, the Government of Malawi (GOM) passed a number of natural resource policies that called for greater decentralization over decision making to lower level administrative units and to communities. Among them were the National Parks and Wildlife Act (1992), the Forestry Act (1997), the Fisheries Management and Conservation Act (1997) and the Local Government Act (1998). All of these promoted co-management (between government agencies and local communities) of natural resources. Committees formed under these acts include Village Natural Resource Management Committees (VNRMCs) (Forestry Department), Water User Associations (Irrigation and Water Department),

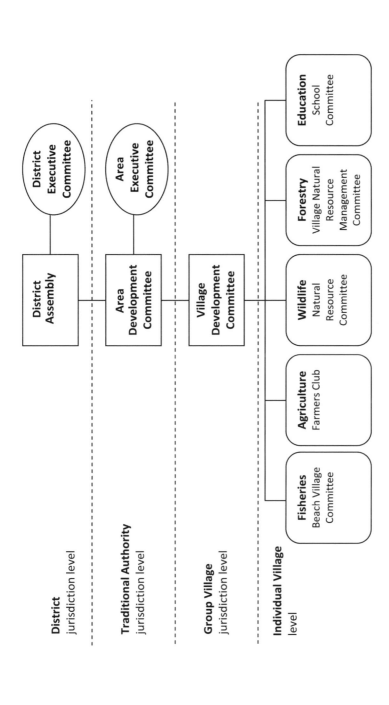

Figure 8.3 The district assembly structure showing the position of natural resource committees

Source: Njaya *et al.*, 2011

Beach Village and River Committees (Fisheries Department) (Figure 8.3) and Waterfowl Committees (Wildlife and Tourism). Of these committees, the Beach Village Committees (BVCs) and Bird Hunting Committees have had greater freedom to issue licences and deliver fines (Figure 8.4). The other committees serve more of an advisory and monitoring role. The degree of authority varies greatly according to involvement of the chiefs. In addition to government institutions and traditional authorities, other institutions involved in management of natural resources include faith-based organizations and non-governmental organizations (NGOs) (Njaya, 2009; Nagoli, 2010).

Decentralization of power to local authorities was seen as a vehicle for poverty reduction as it would allow for better service delivery to Malawians, strengthen democratic institutions and improve participation at the local level (Chiweza, 2010). However, according to Chiweza (2010), the major problem with the decentralization process in Malawi has been a lack of political will and commitment. Indeed, the government continues to postpone local government elections. The stalled decentralization process in Malawi has therefore affected the composition of the district councils (with no elected local representatives) and hence their effectiveness in facilitating approval and implementation of the Lake Chilwa Wetland Management Plan. The institutional arrangements for planning activities and managing development and resources in the basin are illustrated in Figure 8.4 (Nagoli, 2010).

In 1995, Lake Chilwa moved to a co-management system in which communities manage access to the fishery and enforce regulations in partnership with local and national government. At first, the co-management system did not work very well as not all of those groups who depended on the fishery for their livelihoods were well represented. These problems were addressed in 2000 with the passage of the Fisheries Conservation and Management (Local Community Participation) Rules of the Fisheries Conservation and Management Act and elections were held for members of the BVC. In Chilwa, the BVCs are united under an umbrella Lake Chilwa Fisheries Association. The BVCs are authorized to issue fishing licences and to fine violators for not following the fishing regulations. BVC members have themselves identified threats to the important fish species found in the lake which include: cutting of lake plants where fish hide; using gauze wire around plant beds in the lake and marsh; using seine nets in January, February and March; using poison in rivers and pools; catching fish using gauze wire and mosquito nets; and using fish traps in the rivers. They proposed a complete ban on these methods to ensure the health of the fishery. However, enforcing such a ban remains a challenge.

While the Lake Chilwa Wetland Management Plan has not been implemented, there have been some notable small successes in community-based management in the fishery. For example, in a survey of the Lake Chilwa Fisheries Association (an umbrella organization of beach village and river committees) and its BVCs, records indicate an income of 2,160,000 Malawi Kwacha (US$13,500) for 2008, with 37 per cent of that income coming from

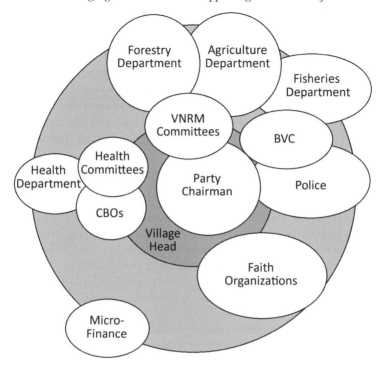

Figure 8.4 Institutional linkages for managing Lake Chilwa basin and wetland
resources

Source: Nagoli, 2010

Note: the inner dark grey circle represents institutions that are based and operate at the local/
community level while the outer light grey circle represents institutions that generally operate
at district and national level.

fines. The rest came from issuing licences. This income was then used for
byelaw enforcement, tributes to the traditional authority (TA), meetings
and assisting sick members of the BVCs. In addition, money was also used
to help with funeral expenses of association members. Thus, in addition to
sustaining the BVCs, income from fines and licences provides important
social services for its members (Wilson, 2009.)

By September 2001, 18 lakeshore communities had created Bird Hunter
Clubs, united under the Bird Hunters Association (Mwayi wa Mbalame) to
better manage and protect the bird population. The clubs drew up regulations controlling the hunting of waterfowl in their areas and have assumed
responsibility for enforcing the regulations, which includes the imposition
of penalties for non-compliance. The Bird Hunters Association was formed
to unite the diverse groups and it decided on the fees that would be charged
for obtaining a bird shooting permit. A total of 26 Bird Sanctuaries were
designated (Wilson, 2004). In the Lake Chilwa Wetland Management Plan,
bird hunting was included under strategies for maintaining biodiversity.

Continuing challenges

Currently, wetland resources are managed by three ministries through four departments (Ministry of Irrigation and Water Development (Water Resources), Ministry of Energy, Natural Resources and Environment (Forestry) and Ministry of Agriculture and Food Security (Land Resources and Fisheries)). These sectoral agencies are represented in the district council with the exception of Forestry Reserves and plantations that are managed centrally by the Forestry Department. Natural resource policies are also designed at the national level.

These existing institutional arrangements are not proving to be very effective for managing wetland resources for a variety of reasons (Lowore and Lowore, 1999; Njaya, 2009; Nagoli, 2010). First, there is a lack of sectoral harmonization with the policies of one sector conflicting with another. The lack of coordinated planning at the local and district level can lead to conflicting messages to local communities (Nagoli, 2010). Second, while some autonomy has been given to the decentralized structures, there is still little opportunity for these units (such as the BVCs) to affect or create policy and regulations. Third, there is also considerable opportunity within the decentralized units for the elite to capture authority and resources. For example, local leaders often use their authority to own beaches, extract rent and control access to resources (Njaya, 2009). Similar observations of elite capture have also been made in irrigation schemes (Kambewa *et al.*, 2007). In the fisheries sector, the existing co-management regime in Lake Chilwa is characterized by lack of participation by fishermen in the formulation of regulations, lack of transparency as most decisions are made by local leaders and the Department of Fisheries, and lack of recognition by fishermen of the role of district councils.

In addition, there has been little institutionalization of natural resources management into the district council's decentralized structure (Lowore and Lowore, 1999; Njaya *et al.*, 2011). Lack of coordination between sectors and actors is largely due to the fact that line agencies such as fisheries, forestry and wildlife have not been fully integrated into one administrative unit at the district level (Chiweza, 2010). Corruption issues have also hampered the implementation of co-management as individuals gain access to resources and circumvent byelaws through bribes (Lowore and Lowore, 1999). Finally, poor participation of communities in co-management activities has impeded success. The root cause of this is thought to be the absence of a functioning decentralized structure and institutionalization of NRM issues into local government structures (Chiweza, 2010).

The management plan developed for the basin in 2001 focused on the three districts in Malawi but neglected to include the Mozambique side of the lake. Due to hydrological and ecological linkages across the whole basin a trans-boundary approach to management is needed. The boundary between the two countries is very porous and there is much movement across it. The Mozambique side of the lake is quite different from the densely populated Malawi side and, with few people, also has more forest cover and less land degradation. Furthermore, the major river running into Lake Chilwa from the Mozambique side, the Mnembo, has better water quality and more species of fish than the rivers on the Malawian side. The soil loss in the Mnembo catchment is significantly lower than any of the river catchments on the Malawi side (Jamu *et al.*, 2006).

There are two major reasons for the exclusion of the Mozambican part of the wetland in the management plan. First, the Lake Chilwa Wetland Management Plan was developed under a bilateral project between Denmark and Malawi. Second, the Lake Chilwa wetland is not among the prioritized list of wetlands for designation as Ramsar sites in Mozambique (Government of Mozambique, 2012). The development of a holistic management plan for Lake Chilwa requires the involvement of Mozambique and the creation of a trans-boundary scheme for joint implementation.

Conclusions

As a Ramsar wetland site that is used extensively by people, the management of Lake Chilwa needs to ensure the wise use of the wetland in order to meet both conservation and development needs. The Lake Chilwa Wetland Management Plan provided a framework for implementing activities that address biodiversity conservation and NRM in the Lake Chilwa wetland. While the plan provided numerous strategies for improving NRM, how to simultaneously improve livelihoods was less defined. Within the basin, both the progress of the Bird Hunters Association and the BVCs has provided an example of how revenue can be generated and resources protected. The Bird Hunters Association has helped to establish waterfowl sanctuaries and develop sustainable utilization plans for waterfowl which has reduced wanton hunting and improved the earnings of hunters who are members of these committees. Similar results have been achieved in the management and conservation of fisheries (Wilson, 2009; Njaya, 2009). Both the BVCs and the Bird Hunters Association struggle with enforcing byelaws as monitoring and patrolling the lake and its wetland is a huge endeavour. While the BVCs have made progress in improving management of the fishery while also raising revenue, more remains to be done to protect the fishery by protecting wetland functions and broader ecosystem integrity (Sarch and Allison, 2000). The importance of peripheral floodplains, marshes and marginal areas of shallow lakes to the overall health of the fishery has been demonstrated and has thus led to the recommendation that management policies for shallow lakes should extend as much as possible to the surrounding land and the wider catchment (Kalk *et al.*, 1979; Njaya, 2001).

In order for the management plan to fully meet biodiversity conservation and livelihoods needs in the Lake Chilwa wetland, it is crucial that local-level natural resource committees are strengthened and made fully functional. With respect to their status in the wetland, there is need to institutionalize these committees into the local-level decentralized structures at the various levels (village, area and district). If these local-level units are strengthened and given more authority and incentives (such as revenue generation), then perhaps the issue of enforcing byelaws will improve.

Institutionalization of NRM committees into the decentralized structures would achieve two main objectives. First, it would ensure clarity of roles and responsibilities of these committees and their relationship with other institutions (political, economic and development) within the decentralized structure thereby improving their functionality. Second, institutionalization of NRM committees would improve the formulation and implementation of management plans at all levels of decentralization because the diverse interests of various sectors (e.g. water, fisheries, agriculture) and institutions (political, economic, etc.) would be taken into account during the development of management plans. This in turn would reduce inter-sectoral conflicts and improve management of trade-offs between various ESS in the wetland.

With respect to livelihood needs, sustainable exploitation plans should include mechanisms and strategies for more productive utilization methods by adding value to harvested products in the wetland. This would include reduction of post-harvest fish losses, in particular through the provision of processing facilities, value addition through packaging and certification of fish products for sale to high-value markets, and more efficient charcoal production methods that reduce the amount of wood required for legal charcoal production. These mechanisms and strategies would ensure that wetland communities derive more income from harvested products while at the same time helping them progress towards sustainable exploitation of wetland resources. Such initiatives are already taking place in the wetland (LCBCCAP, 2011). However, considerable effort still needs to be made to design new opportunities to improve livelihoods in the basin while also sustainably managing the natural resources base. While livelihoods in the basin are now almost entirely dependent on natural resources, establishing small industries that add value to local products could create jobs that enable basin residents to reduce their dependency on farming and fishing.

Population growth in recent decades has increased the pressures on the wetland. In light of continued population growth, it is anticipated that some levels of resource use may be difficult to sustain in the long term. Consequently, planning the use of wetland resources is a priority if the diverse benefits that local communities presently receive are to be sustained. In line with conclusions drawn from analyses elsewhere (Wood and van Halsema, 2008), the integration of agricultural, forestry and fisheries management practices is seen as a prerequisite for sustainable wetland development in many African wetlands. Due to the inter-linkages between activities and resource use in the catchment and the Lake Chilwa ecosystem, increasing emphasis on the management of the wetland must take place within the wider context of the economic and ecological system in which it is situated. Stronger linkages are required between upstream and downstream governance units, as well as greater awareness of the positive and negative impacts of sectoral plans on different resource systems within the basin. Ultimately, the communities who live within the basin and depend on its resources need to be more effectively involved in the design of any management plans, and mechanisms should be put in place to enable them to more effectively improve their livelihoods while also serving as stewards of Lake Chilwa's natural resources and biodiversity.

References

Allison, E. H. and Mvula, P. M. (2002) 'Fishing livelihoods and fisheries management in Malawi', *LADDER Working Paper no. 22*, DFID, London.

Balarin, J. D. (1985) *Aquaculture practices in Africa: systems in use and species cultivated – A status report*, Proceedings, Consultative Workshop on Village Level Aquaculture Development in Africa, Freetown, Commonwealth Secretariat, London.

Beusekom, M. (2011) 'The influence of interactions between traditional and new institutions, water availability as well as nutrient and labour investments on rice yields and economic productivity of farmers during winter season of 2010', MSc thesis, Wageningen Agricultural University, the Netherlands.

Bhima, R. (2006) 'Subsistence use of water birds at Lake Chilwa, Malawi', in G. C. Boere, C. A. Galbraith and D. A. Stroud (eds) *Water birds around the world*, The Stationary Office, Edinburgh, pp. 244–256.

Calder, I. R., Hall, R. L., Bastablea, H. G., Gunstona, H. M., Shela, O., Chirwa, A. and Kafundu, R. (1995) 'The impact of land use change on water resources in sub-Saharan Africa: a modelling study of Lake Malawi', *Journal of Hydrology*, vol. 170, pp. 123–135.

Chavula, G. M. S. (1999) 'The evaluation of the present and potential water resources management for the Lake Chilwa Basin including water resources monitoring', *State of the Environment Study no. 3*, Lake Chilwa Wetland Project, Zomba, Malawi.

Cheyo, D. (1999) *Geohazards around the Michesi and Zomba area*, proceedings of the symposium on natural geological hazards in Southern Malawi, Zomba, 27th–28th July 1999.

Chimatiro, S. K. and Vitsitsi, E. G. (1997) 'Impact of different land-use activities in the catchment on small-scale fish farming in the southern region of Malawi: coordination with other sectors', in K. Remane (ed.) *African Inland Fisheries: Aquaculture and the Environment*, Fishing News Books, Oxford.

Chiwaula, L. and Chaweza, R. (2011) *A firewood and charcoal value chain analysis study for the Lake Chilwa Basin*, Lake Chilwa Climate Change and Adaptation Programme, Zomba, Malawi.

Chiweza, A. L. (2010) *A review of the Malawi decentralization process: lessons from selected districts*, a joint study of the Ministry of Local Government and Rural Development and Concern Universal, Lilongwe, Malawi.

Environmental Affairs Department (1999) *State of environment report*, Environmental Affairs Department, Lilongwe, Malawi.

Environmental Affairs Department (2000) *Lake Chilwa wetland state of the environment*, Environmental Affairs Department, Lilongwe, Malawi.

Environmental Affairs Department (2001a) *Lake Chilwa Wetland Management Plan*, Environmental Affairs Department, Lilongwe, Malawi.

Environmental Affairs Department (2001b) *National Environmental Action Plan*, vol. I, Environmental Affairs Department, Lilongwe, Malawi.

Furse, M. T., Kirk, R. G., Morgan, P. R. and Tweddle, D. (1979) 'Fishes: distribution and biology in relation to changes', in M. Kalk, A. J. McLachlan and C. Howard-Wilson (eds) Lake Chilwa: studies of change in a tropical ecosystem, *Monographiae Biologicae*, vol. 35, Dr. W Junk, The Hague, Boston and London, pp. 175–208.

Government of Malawi (2008) *Fisheries Department Annual Report*, Department of Fisheries, Lilongwe.

Government of Malawi (2010) *Fisheries Department Annual Report*, Department of Fisheries, Lilongwe, Malawi.

Government of Malawi and the International Food and Policy Research Institute (IFPRI) (2002) *Atlas of social statistics*, National Statistics Office, Zomba, Malawi.

Government of Mozambique (2012) National report on the implementation of the Ramsar convention on wetlands: National Reports to be submitted to the 11th Meeting of the Conference of the Contracting Parties, Romania, June 2012, www.ramsar.org/pdf/cop11/nr/cop11-nr-mozambique.pdf (accessed on 15 May 2012).

Howard-Williams, C. and Walker, B. H. (1974) 'The vegetation of a tropical African lake: classification and ordination of the vegetation of Lake Chilwa (Malawi)', *Journal of Ecology*, vol. 62, no. 3, pp. 831–854.

Intergovernmental Panel on Climate Change (IPCC) (2007) *Synthesis Report, Contribution of Working Groups I, II and III to the Fourth Assessment Report of the Intergovernmental Panel on Climate Change*, IPCC, Geneva, p. 104.

Jamu, D. M., Banda, M., Njaya, F. and Hecky, R. (2011) 'Challenges to the sustainable management of Malawi lakes', *Journal of Great Lakes Research*, vol. 37, supp. 1, pp. 3–14.

Jamu, D. M., Chimphamba, J. B. and Brummett, R. E. (2003) 'Land use and cover changes in the Likangala catchment of Lake Chilwa basin, Malawi: implications for managing a tropical wetland', *African Journal of Aquatic Science*, vol. 128, no. 2, pp. 119–132.

Jamu, D. M., Delaney, L. M. and Campbell, C. E. (2006) 'Transboundary management plan for the Lake Chilwa catchment area', www.aucc.ca/_pdf/francais/programs/colloquium/1b-Jamu-Campbell.pdf (unpublished paper), (accessed 7 December 2011).

Kabwazi, H. H. and Wilson, J. G. M. (1998) 'The fishery of Lake Chilwa', in K. van Zegeren and M. P. Munyenyembe (eds) *The Lake Chilwa Environment: A report of the 1996 Ramsar Site Study*, Department of Biology, Chancellor College, Zomba, Malawi.

Kalk, M., McLachlan, A. and Howard-Williams, C. (eds) (1979) 'Lake Chilwa studies of change in a tropical ecosystem', *Monographiae Biologicae*, vol. 35, pp. 17–227.

Kambewa, D. (2005) *Access to and monopoly over wetlands in Malawi*, proceedings of international workshop on 'Africa Water Laws', Johannesburg, South Africa.

Kambewa, P., Mataya, B., Sichinga, K. and Johnson, T. (2007) 'Charcoal: the reality – a study of charcoal consumption, trade and production in Malawi', *Small and Medium Forestry Enterprise Series no. 21*, International Institute for Environment and Development, London.

LCBCCAP (2011) *Lake Chilwa basin climate change adaptation programme annual report*, LEAD-SEA, Zomba, Malawi.

Lowore, J. and Lowore, J. (1999) *Community management of natural resources in Malawi, State of Environment report no. 10*, Lake Chilwa Wetland and Catchment Programme, Zomba, Malawi.

Macuiane, M. A., Kaunda, E. K. and Jamu, D. (2011) 'Seasonal dynamics of physico-chemical characteristics and biological responses of Lake Chilwa, Southern Africa', *Journal of Great Lakes Research*, vol. 37, supp. 1, pp. 75–82.

McCartney, M., Rebelo, L. M., Senaratna Sellamuttu, S. and de Silva, S. (2010) 'Wetlands, agriculture and poverty reduction', *IWMI Research Report 137*, International Water Management Institute, Colombo, Sri Lanka.

Mulwafu, W. O. (2000) *Conflicts over water use in Malawi: a socio-economic study of water resource management along the Likangala River in Zomba District*, proceedings of 1st WARFSA/WaterNet Symposium: Sustainable use of water resources 2000, Maputo, Mozambique.

Munthali, S. M. (1997) 'Dwindling food fish species and fishers' preference: problems of conserving Lake Malawi biodiversity', *Biodiversity and Conservation*, vol. 6, no. 2, pp. 253–261.

Nagoli, J. (2010) *Livelihoods analysis report for the Lake Chilwa Basin*, Lake Chilwa Basin Climate Change and Adaptation Programme, Zomba, Malawi.

Njaya, F., Snyder, K. A., Jamu, D., Wilson, J., Howard-Williams, C., Andrew, N. and Allison, E. (2011) 'The natural history and fisheries ecology of Lake Chilwa, southern Malawi', *Journal of Great Lakes Research*, vol. 31, supp. 1, pp. 15–25.

Njaya, F. J. (2001) 'Review of management measures for Lake Chilwa, Malawi: final project. Fisheries Training Programme', United Nations University, http://www.unuftp.is/static/fellows/document/fridayprf.pdf (accessed 22 September 2010).

Njaya, F. J. (2007) 'Governance challenges for the implementation of fisheries co-management: experiences from Malawi', *International Journal of the Commons*, vol. 1, no. 1, pp. 123–139.

Njaya, F. J. (2009) 'The Lake Chilwa household strategies in response to water level changes: migration, conflicts and co-management', PhD thesis, University of the Western Cape, Republic of South Africa.

Njaya, F. J., Dondaa, S. and Bénéb, C. (2012) 'Analysis of power in fisheries co-management: Experiences from Malawi', *Society and Natural Resources: An International Journal*, vol. 25, no. 7, pp. 652–666.

NSO (2008) *Malawi housing and population census final report*, National Statistical Office, Zomba, Malawi.

Poschinger, A., Cheyo, D. and Mwenelupembe, J. (1998) 'Geohazards in the Michesi and Zomba Mountain Areas in Malawi', *Technical cooperation report*, no. 96, 20915.5, Geological Survey, Zomba.

Rebelo, L. M., McCartney, M. P. and Finlayson, M. C. (2011) 'The application of geospatial analyses to support an integrated study into the ecological character and sustainable use of Lake Chilwa', *Journal of Great Lakes Research*, vol. 37, supp. 1, pp. 83–92.

Sarch, M. T. and Allison, E. A. (2000) 'Fluctuating fisheries in Africa's inland waters: well adapted livelihoods, maladapted management', www.oregonstate.edu/dept/IIFET/2000/papers/sarch.pdf (accessed on 24 September 2012).

Schuijt, K. D. (1999) *Economic valuation of the Lake Chilwa wetland*, State of the environment study no. 18, Lake Chilwa Wetland Project, Zomba, Malawi.

Schuijt, K. D. (2005) 'Economic consequences of wetland degradation for local populations in Africa', *Ecological Economics*, vol. 53, iss. 2, pp. 177–190.

Schuijt, K. D. and Brander, L. (2004) *The economic values of the world's wetlands*, World Wide Fund for Nature (WWF), Gland, Switzerland.

Senaratna Sellamuttu, S., de Silva, S., Nguyen-Khoa, S. and Samarakoon, J. (2008) *Good practices and lessons learned in integrating ecosystem conservation and poverty reduction objectives in wetlands*, International Water Management Institute, Colombo; Wetlands International, Wageningen, the Netherlands.

van Zwieten, P. A. M. and Njaya, F. (2003) 'Environmental variability, effort development, and the regenerative capacity of fish stocks in Lake Chilwa, Malawi', in E. Jul-Larsen, J. R. Kolding, R. Overa, N. J. Raakjaer and P. A. M. van Zwieten (eds) *Management, co-management or no management: major dilemmas in southern Africa freshwater fisheries 2: Case studies*, FAO Fisheries Technical Paper 426/2, Rome, pp. 100–131.

Water Department (2010) *Water Department Annual Report*, Ministry of Irrigation and Water Development, Lilongwe, Malawi.

Wilson, J. G. M. (2004) *An inventory of existing knowledge of bird hunting and community management of birds on Lake Chilwa*, unpublished report.

Wilson, J. G. M. (2009) *Evaluation of participatory fisheries management in Malawi*, unpublished report.

Wood, A. and van Halsema, G. E. (2008) *Scoping agriculture-wetland interactions: towards a sustainable multiple response strategy*, FAO Water Report 33, Food and Agriculture Organization of the United Nations, Rome.

Zimba, G. M. and Kaunda, C. C. (1999) *Health hazards of the Lake Chilwa wetland and catchment*, State of the Environment Study no. 19, Lake Chilwa Wetland Project, Zomba, Malawi.

9 Agriculture, livelihoods and fadama restoration in northern Nigeria

Adamu I. Tanko

Summary

The ecosystem services (ESS) of the fadama wetlands in northern Nigeria have contributed to the livelihoods of people for many centuries. The potential of these wetlands for more intensive utilization has been recognized by the communities and by government agencies. In particular, the droughts of the 1970s encouraged both the expansion of small-scale irrigation with motorised pumps as well as formal irrigation scheme development with storage dams in the upper reaches of several rivers. While crop production was increased, there were negative impacts upon the fadamas and the rivers as a result of the alteration of river flow by the storage dams. These developments seriously impacted upon the livelihoods and living conditions of communities, with fadama cultivation and fishing catches reduced. Over the decade or so after the dams were constructed it began to be recognized that the livelihood losses in the fadamas needed to be addressed as these areas could be more productive than the areas under formal irrigation schemes. This chapter reviews the situation in the Hadejia-Jama'are-Komadugu-Yobe basin (HJKYB) where a Trust Fund (TF) has been established to apply primarily technical solutions to address the major problems caused by the alteration of the river regimes. The TF activities have tried to re-establish the range of ESS that used to be available for the communities in these wetlands. While the initial survey shows that there were major livelihood benefits from these measures, the financial costs have been considerable and further measures may be needed to fully achieve the rehabilitation sought.

Wetlands and river basin development

Wetlands are often reported to be fragile areas that can easily be destroyed by inappropriate interventions and development in a river basin (Ramsar, 2010). While fragility varies from one wetland site to another, there are many examples of degraded wetlands, not least in the Montreux record of endangered Ramsar sites which that convention maintains (Ramsar, 2012). Rehabilitation of degraded wetlands to their 'natural' state is a major task which is costly and time consuming, and difficult to achieve (Erwin, 2009). In particular, dams that are built to control river flows for irrigation or hydropower production have often been a major cause of disruption in downstream wetland areas. Such dams are likely to be an increasing phenomenon in the coming years to support economic development and as an adaption strategy to address the hydrological impacts of climate change (Acreman, 2012).

Over the last three decades increased attention has been given to the impact of dams on stream flows and downstream wetlands (WCD, 2000; Moore *et al.*, 2010), and guidance has been produced to help manage dams better so that they impact less on river regimes and their management gives greater consideration to the original river flows (McCartney, 2009). In particular, there has been the development of decision support systems that seek to provide dam managers with information that will allow them to re-optimize dam releases in the light of the original and new demands for water that have developed. Such measures usually seek to replicate the original river flow to some extent and provide a pattern of flow in rivers affected by dams which will help maintain ESS and meet the need of environmental functioning and provisioning services downstream (McCartney and King, 2011). This is often termed maintaining environmental flows and involves trade-offs between different interest groups (Acreman and McCartney, 2000).

Such management of dams and river flows is very much in line with the ideas presented by the Millennium Ecosystem Assessment (MA) and the Guidelines on Agriculture and Wetlands Interaction (GAWI) studies which both stressed the need to keep a balance of ESS in wetlands for them to be used sustainably. Dam management can also be supported by other measures to improve the efficiency of water use, such as through a systems approach to coordinate water use, as with integrated water resources management and integrated river basin development, and by technical measures such as drip irrigation to reduce the demand for water (Newson, 1997; Lenton and Muller, 2009). The impact of dam construction upon the fadama wetlands in northern Nigeria is discussed in this chapter with a focus upon:

- the disruption to ESS in wetlands caused by insensitive top-down planning of the upstream dams and irrigation development;
- the impacts of these changes in ESS upon the livelihoods of the communities and poverty in this area; and
- the experience to date with remedial measures that try to address these issues and the livelihood improvements that have been achieved.

After a review of the importance of fadamas and various programmes for their development, this chapter focuses on the situation in the HJKYB and in particular the remedial interventions of the Hadejia-Jama'are-Komadugu-Yobe basin Trust Fund (HJKYBTF). This shows that the restoration strategies have had some effects on the ESS and especially the livelihood income of farmers using the fadamas. The chapter is informed by field-based studies in the HJKYB in 2007 and 2010 (Kwaghe *et al.*, 2008; Tanko *et al.*, 2010). The surveyed villages were located in the states of Kano, Bauchi, Jigawa, Yobe and Borno (Figure 9.1). The finding of the surveys and the assessment of the effectiveness of the restoration initiatives allow conclusions to be drawn about ways to ensure sustainable wetland management of the fadamas in this river basin.

Fadamas in northern Nigeria

Fadamas are low-lying, seasonal floodplains that are found in stream and river valleys across the Guinea, Sudan and Sahel savannas of northern Nigeria, from Sokoto in the west to Maiduguri in the east (Kolawole *et al.*, 1994). They are found at the edge of low terraces in the 'alluvial channel complexes' (Ahmed, 1987) which have been created by the major river systems. These complexes consist of both old (inactive) and current (active) flats. The old alluvial flats are abandoned floodplains and back

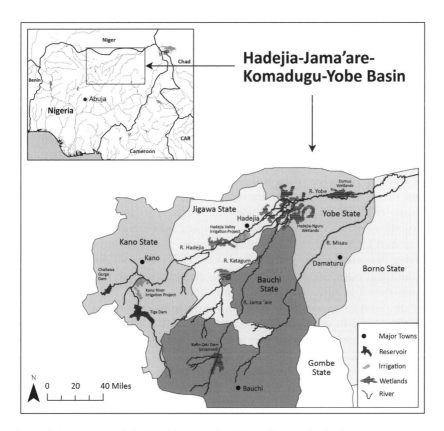

Figure 9.1 Location of the Hadejia-Jama'are-Komadugu-Yobe basin

swamps originating from the migration of the rivers over very extensive channel-cum-floodplains. Active channels are sandy, flat-bedded and steep-sided. Examples of alluvial channels several kilometres wide are found in the Bunsuru, Karaduwa, Gada and Tagwai Rivers, which all flow north-west-wards and eventually join the Sokoto-Rima river system.

Given their hydrological and geomorphological characteristics, the fadamas provide a range of ESS for the local communities which are well recognized in this area, especially regulating and support services (Polet and Thompson, 1996). In particular, fadamas are noted for the role they play in moderating peak flows and flooding during the wet season, as well as helping maintain stream flow during the dry season (Hollis *et al.*, 1993a). In this sense they act very much like typical floodplains (Bullock and Acreman, 2003). Further, they are reported to play a major role in ground water recharge (Hollis *et al.*, 1993a), the control of erosion and the local amelioration of climate extremes (Olofin, 1991).

Fadamas have fertile and friable soils (Tanko, 1999) and are underlain by shallow aquifers so cultivation is possible not only during the flood and flood recession time (for rice), but also during the dry season (Scoones *et al.*, 1996). Given that rainfall in this part of Nigeria is mostly under 1,000 mm and can be less than 500 mm in the north-east, with a rainy season of only four months, fadamas are highly valued, especially in the dry season. Traditionally they have provided dry season grazing especially for the migrating cattle herds of the Fulani from the north, while for the resi-dent Hausa population they have provided valuable opportunities for flood recession and small-scale irrigated agriculture. In the flood season and in the wetter parts of the fadama, fishing has been a major activity (Hollis *et al.*, 1993b). Throughout northern Nigeria, fadamas are the major source of domestic water for rural communities and some larger settlements. Clay for pot making, reeds for crafts and natural medicine have also been obtained from them, although use of these materials has declined in recent years (Tanko *et al.*, 2010). The traditional patterns of socio-economic activities in the floodplains were and remain complex reflecting the diversity of the water and land combinations in the fadama, as well as location relative to market (Thomas, 1996; Tarhule and Woo, 1997). Overall, some 35 per cent of the rural population in northern Nigeria make some use of fadamas to support their livelihoods (World Bank, 2003).

Cultivated fadamas total between 1.5 to 2.5 million ha out of the 33 million ha of land in cultivation in northern Nigeria (FAO, 1997). Crops are grown for domestic consumption as well as for sale. Access to market – in the towns, cities and industrial complexes – is a key variable affecting the latter. Cash crops include rice, cowpeas and sesame, in addition to varieties of valuable market-garden crops such as tomatoes, green vegetables (spinach, cabbage, lettuce), peppers, egg-plants, onions, sweet and Irish potatoes. The fadamas also accommodate livestock of different types including cattle, small rumi-nants and poultry. In some localities farmers may have one predominant type of livestock or depend on livestock keeping as the core of their economy. Overall, the fadamas make an important contribution to food security which is a pronounced issue in rural northern Nigeria due to unfavourable environmental and macroeconomic conditions (USAID, 2007).

The situation in the fadamas must be seen within the wider environmental context of growing pressures on the land resources of northern Nigeria, with land degradation and catchment use not just affecting rainfed cropping but also impacting on the conditions in the wetlands. In particular, the expansion and intensification of agriculture in this part of Nigeria (Mortimore *et al.*, 1990; Mortimore, 1993) has seen major changes in land cover and much of the non-farm vegetation types have been highly degraded (Blench, 2004). Between 1976/1978 and 1993/1995 the area under agriculture increased by 17 per cent to 140,000 km^2 while forest, woodland, grassland and shrubs declined by 26 per cent to 71,450 km^2. In addition swamp forest declined by more than 40 per cent to only 7,000 km^2 while degraded land has increased six fold to 6,565 km^2. This has had consequences for water resources and in turn for the fadamas due to reduced dry season river flow, siltation, flooding and erosion. In addition to these problems, fadamas are becoming susceptible to salinization and alkanization due to fertilizer use and disrupted water flows (Tanko, 1999).

Fadama development in northern Nigeria

Nigeria is a largely agrarian society; agriculture employs 77 per cent of the working population and contributes 42 per cent to the gross domestic product (GDP). Low agricultural productivity is seen as an underlying cause of poverty and as a result in recent decades the government has invested significantly to increase production (UNDP, 1998).

The accelerated agricultural development of the fadamas of northern Nigeria began late in the 1960s with informal efforts by individual farmers in response to recurring drought, which affected over 15 million people and 65 per cent of northern Nigeria. This development was also stimulated by the increased demand for food and water resulting from population increase and urbanization, as well as specific government policies to reduce food imports through dry season irrigation (Kimmage, 1991; Olomoda, 2003). At this time farmers began to irrigate their farm plots adjacent to the river channels for a few months after the rainy season (in October and November) in order to supplement the rainfall and residual flood water available for cropping. In particular, during the 1980s there was a massive increase in irrigated wheat cultivation by individual farmers in the fadamas with the area of this crop expanding from 200–300 ha in 1983 to 47,000 ha by 1989/1990 (Hollis *et al.*, 1993b).

In time, the drought and food security situation forced the state and federal governments to respond, which involved the construction of dams on a number of rivers. In the Kano Region alone, 17 dams were constructed between 1969 and 1976, while across the whole of northern Nigeria more than 46 have been built since 1969. These dams were built in the upper sections of the rivers and impounded between 5.3 million m³ (at Mairuwa in the Sokoto-Rima basin) and 11,500 million m³ (Kainji in the Niger river basin) (JICA, 1995).

While these dams are primarily for irrigation, they also contribute to urban water supply and to a lesser degree generate hydro-electricity (Tanko, 2010). The formal irrigation schemes with their canal distribution systems are mostly found in the upper river fadamas, and cover between 5,000 ha and 25,000 ha, with the number of beneficiary farmers per scheme varying from 850 to over 10,000. The irrigation water has been used to supplement rainy season cultivation of rice, as well as to provide a secure source of water for dry season cultivation of wheat and vegetables.

While the dams and irrigation facilities increased agricultural output in the areas within the schemes, a range of serious problems began to appear in other parts of the river systems, especially the downstream fadamas. In particular the damming of the rivers reduced the extent of the flood in the wet season (Olofin, 1987), while the provision of sediment onto the fadamas, which was one way in which their fertility was maintained, was also reduced (Adewunmi, 1999). Lower river flows in the rivers also led to lower water velocities which in turn resulted in increased sediment deposition within the river channels (Bird and Tanko, 2004). The lower flows and sediment build up allowed some plants to proliferate, especially *typha* (a perennial herbaceous plant), which in turn had a cumulative effect, slowing river flows further and increasing sedimentation (IUCN/HNWCP, 1999). In turn, this led to altered river courses and to some areas suffering permanent inundation, while other areas remained unflooded in the wet season (Tanko, 1999). While channel variations had often caused some of these problems in the past and required community efforts to redirect streams and remove sediment, the scale of these problems was vastly increased after the dams began to alter the flow regimes in the rivers (Kwaghe *et al.*, 2008).

While some people, mostly in the irrigation schemes, have benefitted from the irrigation water from the dams, the overall result for the fadama-using population was negative with the loss of farmland and grazing areas, and declining fish yields, leading to increased poverty and out-migration (Binns, 1995). In addition, tension between the different fadama stakeholders increased with the altered flooding regimes making shared use of these areas more difficult (JEWEL, 2003). Some studies indicate that despite the fadama development activities, poverty in the area increased several fold between 1993 and 2006 (Olukesusi, 2011).

The serious problems associated with the damming of the main rivers in northern Nigeria eventually led to an enhanced recognition of the value and potential of the fadamas, which in turn led to efforts to better use these areas despite the altered river regimes. In particular, a series of Fadama Development Projects were launched in the early 1990s to improve production from these areas. Initially, the focus of these measures was on using the fadama as they were at that time, although they later developed some remedial actions to cope with the impacts of the dams. The Fadama Development Programme, which was initiated by the Federal Government of Nigeria (Box 9.1), had support from the World Bank and the African Development Bank (ADB), and was seen as a major contribution to the country's rural development challenges (UNDP, 1998).

Box 9.1 The National Fadama Development Programme (NFDP) (CADP, 2007)

Implementation of the National Fadama Development Programme (NFDP) began in 1993 with the launching of Fadama I in selected parts of northern Nigeria. This had the aim of developing small-scale irrigation through the extraction of shallow groundwater with low-cost, petrol-driven pumps. In total over 40,000 tubewells were installed and together with the pumps they made irrigation more reliable for a longer period (Tarhule and Woo, 1997). The adoption of this technology enabled farmers to increase production by more than 300 per cent in some cases. The Federal Government, impressed by the achievements of Fadama I, approached the African Development Fund (ADF) of the ADB for support in expanding the project by adopting a Community-Driven Development (CDD) approach with extensive participation of stakeholders. This Fadama II project even included some parts of southern Nigeria. The CDD approach emphasized poverty reduction, private sector leadership and beneficiary participation. Fadama II commenced in January 2004 and lasted six years. It is credited with increasing the income of farmers and offering new employment opportunities that reduced poverty although the precise impact has not been assessed (CADP, 2007).

In order to further strengthen agricultural production, processing and marketing, Fadama III was launched in 2010. This focuses on small and medium-scale commercial farmers and agro-processors with the aim of contributing to poverty reduction, increased food security and the achievement of other key Millennium Development Goals (MDGs). Fadama III also directly and indirectly supports subsistence and poor farmers with information, skills, technology, group organization and business opportunities that will allow them to pursue micro-enterprises, self-employment or other opportunities in commercial agriculture, as well as assist them through training to become employed in market-chain activities. Positive impacts on social and gender development are expected through:

- expanding opportunities for the poor and women to engage in commercial activities;
- reducing the vulnerability of disadvantaged groups arising from the commercialization of agriculture; and
- enhancing capabilities to engage directly in or benefit indirectly from commercial agriculture.

Investment in local infrastructure in the form of feeder roads, appropriate agricultural technologies, market information and agribusiness and/or product improvement technology is to be provided.
   Key lessons about the programme so far include:

- Higher income levels for participating farmers. For instance, Adegbite *et al.* (2007), using the gross margin analysis at various scales of operation, revealed that Fadama II-participating farmers with farm sizes of between 0.1 and 0.99 ha had margins of N97,347/ha (about US$610) as compared to the non-participating farmers who had N17,721/ha (about US$110). For farm sizes between 1.00 and 1.99 ha participating farmers had margins of N117,257/ha (about US$730) as compared to N31,434/ha (about US$200) for the non-participating counterparts.
- Improved access of the poor to basic services, opportunities for social advancement and participation in the development process.
- Provision of demand-driven assistance, results-focused support and flexible implementation modalities.
- Encouragement of community participation with social inclusion in overall project implementation.
- Provision of services and support in a way that is transparent and accountable.

Development and restoration in the Hadejia-Jama'are-Komadugu-Yobe basin

The impacts of dam development

The HJKYB is in a semi-arid sub-catchment of the much larger Lake Chad basin and the Sudan-Sahel zone of north-eastern Nigeria. It has a combined catchment of approximately 84,000 km^2 (Figure 9.1) and includes 45 per cent of the fadamas of northern Nigeria (CADP, 2007). The total population in the basin is just over 20 million, of whom over 15 million are rural dwellers engaged in agriculture, fishing, livestock keeping and the collection of natural products (Abba *et al.*, 2006). About nine million people, 60 per cent of the rural dwellers in the area, are thought to depend on fadama resources for their livelihoods (Abba *et al.*, 2006). The populations in the major cities and towns like Kano, Bauchi and Maiduguri, as well as most of the rural settlements, depend on water from the fadama and the rivers that flow through them.

The HJKYB has had two major dams built, the Tiga and Chalawa Gorge dams. Both are located in the upstream part of the basin. The Tiga Dam was completed on the Hadejia River in 1974 with an active reservoir capacity of 1,400 million m³. In 1992, the spillway level was lowered to preserve the structural stability of the dam, which resulted in a 31 per cent reduction in its storage capacity (IUCN/NIWRMC, 2011). While the water is used primarily in the Kano Irrigation Project (22,000 ha), the dam also provides water for municipal and industrial purposes in metropolitan Kano. According to Tanko (1996), before the construction of the Tiga Dam, the Hadejia river system used to have relatively strong stream flow during the months of June to October and this accounted for 98–99 per cent of the annual flow of the Yobe River. The Challawa Gorge Dam on the Challawa River, a tributary of the Hadejia River, was completed in 1992 with a reservoir capacity of 972 million m³ (Tanko, 2010). Water from this reservoir, which was originally intended to supplement water provision for the Kano River Project Phase II, now supplements the Tiga supplies to the Kano metropolis and into the Hadejia River for subsequent storage behind the Hadejia Barrage which also supplies the Hadejia Valley Irrigation Project (HVIP) further downstream. The dams and large-scale water management in the basin are currently managed by the Hadejia-Jama'are River Basin Development Authority (HJRBDA) which was established in 1979.

Since 1974, dry-season water releases from the reservoirs have modified the dry-season flow from hitherto zero to almost a perennial regime (IUCN/NIWRMC, 2011). This dry-season flow is in part responsible for the development of invasive aquatic weed infestation, especially *typha*, and other negative impacts in the river system downstream of Gashua (Oyebande, 2003). Conversely, lower flows in the wet season have worsened the sedimentation problems which in turn have aggravated the growth of the invasive weeds and led to blockages along the old Hadejia river channel. This in turn progressively reduced, and in the end prevented, its contribution of water to the Komadugu-Yobe which impacted drastically on the ecology of the basin. This led to a decline in the extent of the wetlands and a decrease in the fish population, with the loss of at least five species (Oyebande, 2003). Overall it was estimated that the area subject to annual flooding had been reduced by 70 per cent or more by 1993, from 300,000 ha to 70,000–90,000 ha (Hollis *et al.*, 1993b). Hence the overall situation at its worst was a gain of 22,000 ha of formally irrigated land subject to two crops a year with the loss of more than 200,000 ha of seasonally inundated wetland with multiple uses. The resulting changes in the pattern of livelihood benefits led to a reassessment of the value of the ESS of the fad-amas (Olofin, 1994; Barbier, 2003) and to proposals for interventions to try to better manage the water resources in these areas and re-establish a balance in the ESS through the river system and especially in the fadama.

The search for practical restoration measures was the result of a long pro-cess of coming to understand this situation and its dynamics, and engage the various stakeholders. This was led by the Nigerian Conservation Foundation, and included internationally supported projects with the International Union for the Conservation of Nature (IUCN) and the Department for International Development (DFID) (IUCN/NIWRMC, 2011). Through these projects and following a consultative process a consensus was reached by all stakeholders in 2005 that an integrated approach to water resources management was needed in the basin (JEWEL, 2003). To progress this, a TF was established in 2006. This not only had the aim of better water resource management in the basin's fadamas, but also identified fadama sustainabil-ity as one of the long-term goals of the TF, with seasonal flooding of the fadamas identified as crucial to the long-term use of these areas by people (and also for the flora and wildlife, especially birds).

The need to address the problems caused by the dams was driven by the increased realization of the agricultural value of the fadamas which were seen as the most productive areas in northern Nigeria (NHI/IUCN, 2007). One report stated: 'the economic value of production from the fadamas is very large, many times greater than that of all the irrigation schemes for which the inflowing rivers are dammed, diverted and their water used' (NHI/IUCN, 2007: 3). Indeed it is recognized that the fadamas produce an agricultural surplus in non-drought years, support a substantial popu-lation at relatively high levels of nutrition and income, and play a critical role in the local economy. It is now agreed that they should never be mis-used. In order to resuscitate the fadamas the TF therefore focused on the construction of structures to help re-establish the pre-dam characteristics of the rivers and fadamas as previously identified by the communities in the Jigawa Enhancement of Wetlands and Livelihood (JEWEL) assessment (JEWEL, 2003).

The restoration programmes of the TF began in November 2007. It was the first attempt to undertake remedial measures in the fadama follow-ing the construction of the dams. There were three major interventions, these being weed/channel clearance, construction of dykes (both earth and concrete) and flood retention gates with a total cost of approximately US$10.6 million. Using the reports of Bird and Tanko (2004) and IUCN/HNWCP (1999) the TF selected the (22) villages for the following interven-tion activities:

- Channel clearance to get rid of typha and silt – this intervention involves the annual removal of typha and excavation of silt by dredging (every 4–5 years) to unblock the channels and re-establish steady water flow into, and within, the fadamas. This intervention was implemented at a number of fadamas, specifically at Rantan, Magujin Idi, Miga, Kafin Hausa, Landa Mada and Gwayo/Gasi. The clearance sought to improve water availability in the fadamas and so make increased cultivation possible.

- Construction of dykes and embankments – these structures were built to protect vulnerable settlements and farmland from flooding. The interventions were designed and implemented at Warawa, Dagu, Faggo, Ariri, Karage and Ituwa fadamas. In these areas, excessive floods and inundation of fadamas had become an annual occurrence and affected the settlements and economic activities. By constructing the dykes and embankments to protect these communities and their fadamas, it was hoped that farming in the protected areas would be improved.

- Construction of flood retention gate structures – this was aimed at conserving and preventing rapid recession of flood water from fadamas and thereby maintaining water in areas where it can support livestock husbandry and fishery activities. Communities where the interventions were designed and implemented were Reni-Kunu, Yusuri, Damasak, Joka Juriye, Dagona, Bulagana Chira'a (Kuwait) and Jiyen. This intervention was expected to improve farming activities based on residual moisture in the fadamas, but also fishing and livestock keeping. The increased income generated was expected to have positive multiplier effects on the general well-being of the people in and around the fadamas.

- Re-optimization of the dam releases was requested by the TF, but the authorities (HJRBDA) have not implemented this or the required studies due to lack of funds.

The villagers were involved in the construction of these measures through the provision of communal labour to support the work undertaken by machinery provided by the TF. They were also given training on how to maintain these measures and operate them – where appropriate – as well as how to benefit from the recreated fadama conditions.

Given the growing overall water deficit in the basin, the various studies have identified a number of other measures that require attention but have yet to be taken up by the TF or other projects. These include:

- re-managing land use in the fadama in ways to reduce water needs;
- improving water efficiency in the irrigation schemes, especially during the rainy season;
- developing a catchment management plan to develop coordinated and a more efficient and integrated approach to water resource management;
- developing a decision support system to help dam management for re-optimization; and
- group formation to try to build up the basis for community input into the management of the basin in a coordinated way (IUCN/NIWRMC, 2011).

Investigating fadama development and restoration

In order to assess the impacts of these interventions two studies were undertaken. A baseline of 17 communities was established in 2007 (Kwaghe *et al.*, 2008) before the various restoration activities were undertaken later that year and completed in 2008. Another survey was carried out in 2010 in the same 17 and another five villages (covering all 22 intervention villages), with the aim of assessing the immediate impact of the projects (Tanko *et al.*, 2010).

These two studies used various Participatory Rural Appraisal (PRA) techniques which were applied through group interviews with traditional leaders, community-based organization (CBO) leaders, youth groups, women's groups and other stakeholders in the surveyed villages where the restoration measures were executed. These studied villages, which each consist of several communities scattered over the village lands, had an estimated total population of over 300,000 people (Kwaghe *et al.*, 2008).

A checklist was used to guide the discussion which investigated the following indices:

- Crop production – types of crops, present production (tons/hectare/annum) of each crop under-irrigated and rain-fed agriculture, including crop residues. Livestock index of the present herds in the intervention areas, e.g. cattle, sheep and goats.
- Fisheries – inventory of fish species present and those that have disappeared, as well as yields at various times in the past.
- Use of water – essentially valuation of the impact of other forms of water use, especially for domestic purposes, before and after the TF's intervention in and around all the communities. The studies were expected to note the current period (months) of water retention at locations of intervention before and after the interventions, as well as variations in expected benefits before and after the interventions.
- Population density and structure, availability and access to social amenities such as potable drinking water, shelter, food and health.

The investigation also explored other conditions (in 2007–2008) before the restoration activities, including seasonal variations in the types of economic activities. In addition, the study attempted to quantify the income from different economic activities within and around the intervention areas, as well as the past effects of floods.

Fadama-based livelihoods

All the studied villages were located downstream of the Tiga and Challawa Dams on the River Hadejia system, as well as downstream of the proposed Kafin-Zaki Dam on the River Jama'are system (see Figure 9.1). Most of the communities within these villages were off tarred roads.

The major environmental concerns reported in most of the villages were water related and varied depending on proximity to the rivers or stream channels. There were problems of long-term inundation due to blocked channels and irregular flooding due to unregulated releases. These two problems affected irrigable and flood/recession plots, and human settlements. Sixteen communities indicated severe water situations in the months of July and August when between 30 per cent and 70 per cent of farm plots were inundated and farmers had to abandon crops in these plots. In six other communities, there were problems of desiccation which especially affected fishing areas. This was a major problem in 20–40 per cent of the fishing areas depending on the village. While the increased floods are clearly a change from previous conditions, there is a major debate on-going over the extent to which natural conditions or the river basin authority – who supervise the dam operation – are responsible for the worst floods.

Invasive weeds in eight communities were also found to be important. Different species were reported. These were locally identified as including typha or *roba, damban bakin kogi, geranya, gwaya-gwaya, aya-aya, tabon angulu, yaryadi, gemun kwalo, tumbin jaki* and *gazar giwa*. Affected communities reported that these had taken over water channels, created blockages, exacerbated floods and caused inundation. In addition to the above, other species had taken over farmland, rendering the plots very difficult to cultivate as they caused waterlogging. This, combined with the high application of chemical fertilizers, had led to problems with soil salinity/alkalinity in some places.

The baseline study confirmed that livelihoods in the fadamas in the basin were overwhelmingly based on the primary production of food and cash crops, along with livestock rearing and fishing. In the fadamas different combinations of crops were grown, including rice, cowpeas and sesame, in addition to a variety of vegetable crops. Moreover, farmers in the fadamas kept cattle, small ruminants and poultry. In some localities farmers had one predominant type of livestock to supplement crop production. Generally, the fadama crops were for domestic use and sale. Table 9.1 summarizes the production range of both food and cash crops in the studied villages before the restoration.

Before the restoration, rice was found to be the highest yielding crop. Livestock in the area were kept primarily for their security value. The major livestock were goats, sheep and cattle and they were normally sold during periods of hardship as a coping strategy.

Fishing was an important livelihood activity in the communities; the majority of households participated either on a full-time or part-time basis. It was reported across the communities that daily catches or the value of the daily catch had fallen drastically when compared to the averages over the previous 10–20 years.

Table 9.1 Agricultural production in the HJKYB fadamas before restoration (2007)

Crops			Livestock			Fishing	
Crop types	Annual production (kg/household)		Types	No./ household		Species (local names)	Sales from daily catches (US$)
	Max	Min		Max	Min		
Sorghum	434.8	83.3	Cattle	32	2	Tarwada,	19–44
Maize	543.5	194.4	Sheep	65	18	Bargi,	
Millet	782.6	222.2	Goat	85	9	Karfasa,	
Rice	1,513.0	222.2				Musko,	
Cowpea	417.4	277.8				Kawara,	
						Tsage,	
						Karaya	

Impacts of the restoration

Channel clearance to remove typha and silt at Rantan, Maguji Idi, Miga, Kafin Hausa, Landa Mada and Gwayo/Gasi

Early in 2008 the TF coordinated action with mobilized communal labour and excavators. This allowed 25 km of channels to be cleared of *typha* and other aquatic vegetation that had caused blockages. As a result, 2,000 ha of land that had been uncultivated in the dry season became suitable for cropping now the flow was assured. Fishermen and farmers who had migrated away from the area began to return. The pilot clearance exercise also served as a demonstration of best practice in terms of catalysing joint action across tiers of government and across sectors that could influence policy. It also served as a practical capacity-building experience for community leaders. Since then agriculture has improved with production and the variety of crops increasing. Table 9.2 shows agricultural production at Tiga, Magujin Idi, Miga, Kafin Hausa and Gwayo/Gasi.

Communities at Rantan, Maguji Idi, Miga, Kafin Hausa, Landa Mada and Gwayo/Gasi indicated that, in addition to traditional crops, there were others cultivated since the restoration. These included groundnut and sorrel. Comparing the average yields of other crops per households, there were increases in almost all crops. For example, the average yields of rice, millet, sorghum and maize per household increased by about 50 per cent. Although this pattern appeared to be similar in the four other communities, rice production remained lower at Gwayo. This was attributed to the persistent blockage of the stream channel which caused desiccation of the fadamas in the area. However, yields of the other three major crops (i.e. millet, sorghum and maize) remained the same over the years.

Table 9.2 Agricultural production in five communities after channels were cleared (2010)

Crops			Livestock			Fishing	
Crop types	Annual production (kg/household)		Types	No./ houshold		Species (local names)	Sales from daily catches (US$)
	Max	Min		Max	Min		
Sorghum	620.0	111.1	Cattle	60	6	*Tarwada,*	13–63
Maize	880.0	277.8	Sheep	200	10	*Bargi,*	
Millet	960.0	277.8	Goat	450	5	*Karfasa,*	
Rice	2,800.0	277.8				*Zari, Musko,*	
Cowpea	600.0	277.8				*Kawara,*	
						Tsage, Karaya,	
						Ganwa, Lulu,	
						Kaunsa,	
						Dundukuri	

The situation of livestock keeping also changed positively in most of the communities. For instance, in one of the communities the total number of cattle in the pre-intervention period was 34 but in 2010 it had gone up to 100. Similarly the total numbers of goats and sheep in the community rose. This was due to increased feed from crop residues and fodder. Farmers in most parts of northern Nigeria reinvest their agricultural gains into livestock keeping.

Fishing in the communities showed some positive changes too. Daily catches/sales were higher, while there were new species identified. For example, in Rantan, where the fishermen created their own fishing ponds, one of the fishermen was reported to have a pond from which he harvested fish catches worth up to N20,000 (about US$125) weekly. As a result, fish farming was becoming popular in the community and most of the young men indicated an interest in fishing.

*Construction of dykes and embankments at Garin Atiku , Dagu, Faggo,
Ariri, Karage, Waek, Tage, Matarar Ganji and Ituwa fadamas*

In all the communities restored through the construction of dykes and embankments, increased crop production was recorded. For example, higher average production yields of rice, millet, sorghum and maize per household were recorded in all the communities (Table 9.3). Comparison of the data on crop production in 2007 and 2010 revealed that irrigated crops such as tomatoes, water melon and chilli pepper were being cultivated. Comparison of fishing activities in 2008 and 2010, however, revealed that the number of

Table 9.3 Agricultural production in seven communities with dykes and embankments (2010)

Crops			Livestock			Fishing	
Crop types	Annual production (kg/household)		Types	No./household		Species (local names)	Sales from daily catches (US$)
	Max	Min		Max	Min		
Sorghum	612.0	64.2	Cattle	150	10	Karfasa,	25–83
Maize	713.4	170.3	Sheep	680	25	Kawara,	
Millet	935.8	305.8	Goat	750	40	Musco, Tsage,	
Rice	1,911.1	675.1				Saro, Tarwada,	
Cowpea	613.4	277.8				Karaya,	
						Burdo, Marde,	
						Lafsa, Tatar,	
						Sawayya,	
						Gaiwa, Bargi,	
						Barya, Jari,	
						Muzuni	

fishermen had fallen from 100 to 20 in the community. This was because of the improvement in the water situation in their fadama which led to most people abandoning fishing for crop production. Those who remained as fishermen, however, had better incomes in 2010 compared to 2008.

Generally, the intervention created improved environmental conditions especially around the agricultural lands. Moreover, as the conditions of the roads improved, communities indicated that they had better accessibility to market and so their farm products fetched better prices.

Construction of flood retention gate structures at Reni-Kunu, Yusuri, Damasak, Joka Juriye, Dagona, Bulagana Chira'a (Kuwait) and Jiyen

The main problems in these communities were twofold. First was the flooding situation on fadama lands. The major reasons for this were the collapse of river banks due to grazing activities and the blockage of the river channels and increased siltation due to the presence of typha. It was estimated that about 70 per cent of the recession/irrigable cropping lands were not usable. Second, there was the quick recession of the flood from the fadama lands back into the river which could have been used for other economic activities if it was retained. Therefore, the intervention applied was the construction of high embankments to safeguard rainfed crops from flooding and the provision of single valve gates to prevent the quick recession of the

Table 9.4 Agricultural production in seven communities with flood retention gates (2010)

Crops			Livestock			Fishing	
Crop types	Annual production (kg/household)		Types	No./household		Species (local names)	Sales from daily catches (US$)
	Max	Min		Max	Min		
Sorghum	587.0	83.3	Cattle	60	6	Tarwada,	19–63
Maize	869.6	194.4	Sheep	200	15	Bargi,	
Millet	1,087.0	222.2	Goat	350	15	Karfasa,	
Rice	2,826.1	222.2				Zari, Musko,	
Cowpea	521.7	277.8				Kawara,	
						Tsage, Karaya,	
						Ganwa, Lulu,	
						Kaunsa,	
						Dundukuri	

fadama flood. Previously the desiccation due to quick recession of the flood had rendered 90 per cent of flooding/recession cropping lands, fishing areas and dry season grazing fadama lands unusable.

In all the communities almost 100 per cent of the members engage in agricultural activities. The majority of the villagers are engaged in crop farming first, followed by fishing, then livestock rearing. After the restoration programmes, the major crops cultivated in all the seven communities showed major improvements (Table 9.4). In particular, the maximum and minimum production of rice, millet and sorghum per household were found to be much higher than in the pre-restoration period. Livestock reared by the communities, in order of importance, were sheep, goats and cattle. The main fish species present were Tarwada, Bargi, Karfasa, Musko, Kawara, Tsage and Karaya.

Balancing ESS and livelihoods

The remedial measures to restore the functioning of the rivers and streams, and their flooding of the fadamas, appear to have had major positive impacts in the selected communities. The above analysis shows that in the 22 villages where interventions were made, there appear to have been considerable improvements in the conditions compared to the situation as a whole before the remedial work was begun. In particular, agricultural production and fishing are providing livelihood benefits that are reportedly similar to, or better than, those before the dams were built. In addition some new crops have been taken up such as water melons and hot peppers. This economic recovery has been dependent upon the partial re-establishment of the pattern of flooding in the fadama similar to that before dam building began. Not only have agriculture and fishing improved but so too has live-stock rearing, although other livelihoods based on reed resources and clay for pottery have not recovered as market conditions have changed (Tanko *et al.*, 2010).

This recovery of the rural economy is reflected in the return of people who had migrated out of their villages after the dams were built, and also in the improved ownership of assets, such as tractors, motor vehicles, grinding mills and water pumps. Other assets now increasingly common in the area are motorcycles, mobile phones, sewing machines and also generators and sprayers. Commercial activities are reported to have seen a resurgence and there are now thriving markets in several parts of the basin and traders are moving into the area to benefit from these opportunities. All of these point to the availability of disposable income. One indication of improved well-being was the increased number of marriages reported in the 22 sampled villages during the marriage season (Kaka) in 2010 when compared to pre-vious years (Tanko *et al.*, 2010).

Other ESS, besides provisioning and regulatory services, have prob-ably also improved in the fadamas of the selected villages. These should include ground water recharge, where the fadamas are thought to be of considerable importance, and a number of support services related to soil structure, nutrient cycling and the reduction in soil acidity (Hollis *et al.*, 1993b; Tanko *et al.*, 2010). Groundwater recharge may have widely felt benefits as there is widespread water extraction for small-scale irrigation and domestic water supply across the area, while reduced soil acidity will improve crop yields. Biodiversity benefits are also probably enhanced now that the vegetation in the rivers is no longer dominated by the *typha* grass, while the return to the previous pattern of flooding should also benefit the migratory bird species.

However, as yet, the problems in the fadama caused by the dams have only been addressed in 22 villages out of more than 80 in the HJKYB. The benefits to date have been achieved by treating the symptoms of the prob-lem, the sediment, *typha* and flooding patterns, rather than addressing the underlying problem of dam management. The latter will require re-optimizing the discharge patterns from the dams so as to make them less intrusive in the river system, and establish a river flow regime similar to that before the dams were built. For this to be achieved there will have to be a trade-off process with the fadama farmers in general benefiting from a less than 100 per cent re-establishment of the traditional patterns of flooding and water flow, while those farmers who have benefitted from the irrigation water from the dams will have to suffer some reduction in the availability of irrigation water. This will require very careful analysis to assess the total range of impacts that will result from a return to a more natural river regime. The overall results might be an increase in the net provisioning benefits across the basin as a whole, while a better mix of ESS would improve the sustainability of wetland functioning and rural livelihoods in the area as a whole. However, given the alterations to date and the adaptations by communities to the present situation, it may be that specific interventions as implemented by the TF may be more appropriate and produce greater net benefits at lower costs.

These developments will have to be part of the on-going discussions about developing a river basin approach to the management of water and land resources in this area which could apply the principles of integrated water resource management and would also improve understanding of the diverse ESS and their livelihood contributions before further interventions are made (HJKYBTF *et al.*, 2007). This approach would help ensure that the interests of all stakeholders are better engaged and the communities are empowered.

Conclusions

The fadama wetlands of northern Nigeria are an important agricultural resource supporting to varying degrees nine million people. They account for between 5 and 8 percent of the cultivated land in this part of Nigeria (FAO, 1997). In addition, they provide other livelihood benefits through fishing, craft activities and the grazing of livestock. The reliable crop production in the fadama, based on residual moisture or irrigation, is especially important given the long dry season and the recurring droughts in this area. Compared to the irrigation schemes for which the rivers were dammed and diverted, the fadamas are estimated to have a much greater economic value (NHI/IUCN, 2007).

Increasing the formal, scheme-based irrigation in these areas was seen by the regional and federal government as a way to improve food production and food security, with river regime alteration, especially through the creation of dams, an acceptable cost. However, this top-down approach failed to consider the full impacts of altering the hydrological regimes in the stream, rivers and wetlands in this area. The negative results on the traditional fadama-based livelihoods were considerable and appear to have more than offset the gains from increased irrigation. The remedial measures that have been introduced in some areas, notably the HJKYB by the HJKYBTF working with the local communities, have helped re-establish to some extent the pre-dam situation with respect to ESS and the livelihoods based on them in 22 specific communities. However, these technical measures are only addressing the symptoms of the altered river flow, the siltation, weed growth, water logging, changed courses and desiccation. What remains essential is the continued negotiation between the HJKYBTF and the HJRBDA to explore the adjusting of the dam releases, through a re-optimization process, supported by building the knowledge for this in a decision support system (McCartney and King 2011). This will help ensure the sustainability of the improvements to date and the achievement of remediation across the basin as a whole. Building the information and support for this will involve many of the measures identified in the numerous reports to date, starting with community organization, the accumulation of appropriate knowledge, the development of monitoring systems and the coordination of actions across the basin through an appropriate basin-wide plan and effective institution (HJKYBTF *et al.*, 2007).

With respect to the search for long-term sustainable wetland use, the lessons from the fadama experience in northern Nigeria are several. First, seasonally flooded wetlands can be of considerable importance in rural economies both for subsistence and rural development and have an important role to play in adapting to droughts and probably also to climate change. Second, when altering the flow of water in the rivers, and especially when withdrawing water from these systems and hence affecting the ESS situation in wetlands, major changes may result that can become cumulative and re-enforcing. They may have negative impacts upon the pattern of ESS and especially the livelihoods that can be obtained from provisioning services. Third, this experience raises questions about the extent to which alteration of the patterns of ESS in wetlands can be made without having major negative impacts and reduced sustainability. Fourth, it is clear that engineering solutions may sometimes be applicable and can improve livelihoods. However, in some situations they may not be effective and reverting to the former situation may not be possible, or even desirable to some groups of stakeholders. Fifth, this case study of fadamas confirms the need to explore how to identify tipping points in the process of change in ESS beyond which reversing change is difficult. A similar question is raised about the extent to which remedial measures can work and how successful they can be and at what cost. Sixth, this experience suggests that given the different interests within the wetlands and with respect to the water resources which support these wetlands, any development measures must consider the trade-offs which need to be made so that overall benefits can be maximized and sustainability ensured. Seventh, this case study shows that there are different bodies of knowledge and wetland interests at the state/government level and at the community level. Both have to be part of addressing the issue of how to achieve a balance of ESS in these wetlands, which will ensure sustainability and will also help address food security and rural poverty. Finally, it is clear that there is a need for a systems view in understanding wetlands and their management, and that this should be linked to a river basin planning approach which can draw on experiences from integrated water resource management.

References

Abba, A., Jibrin, J. M., Suleiman, A. and Mustapha, A. (2006) *Improving Land and Water Resources Management in the Komadugu Yobe Basin: Socio-Economic and Environmental Study*, unpublished project report submitted to FMAWR-IUCN-NCF (Federal Ministry of Agriculture and Water Resources, International Union for the Conservation of Nature, Nigeria Conservation Foundation), Komadugu Yobe Basin Project.

Acreman, M. C. (2012) *Wetlands and Water Storage: Current and Future Trends and Issues*, Ramsar Scientific and Technical Briefing Note no. 2, Ramsar Convention Secretariat, Gland.

Acreman, M. C. and McCartney, M. P. (2000) 'Framework guidelines for managed flood releases from reservoirs to maintain downstream ecosystems and dependent livelihoods', *International Workshop on Development and Management of Floodplains and Wetlands*, Bejing, China, 5–8 September, pp. 155–164.

Adegbite, D. A., Adubi, K. O., Oloruntoba, A., Oyekunle, O. and Sobanke, S.B. (2007) 'Impact of National Fadama Development Project II on small-scale farmers' income in Ogun State, Nigeria: implications for financial support to farmers,' *International Journal of Agricultural Sciences, Science, Environment and Technology*, Series C, vol. 2, no. 1, pp. 110–130.

Adewunmi, M. O. (1999) 'Small scale irrigation management in Nigeria: a study of fadama farming in Kwara State, Nigeria', *Agrosearch*, vol. 5, nos 1 and 2, pp. 1–6.

Ahmed, K. (1987) 'Erosion hazard assessment in the Savanna: the Hadejia-Jama'are River Basin', in Mortimore, M., Olofin, E. A., Cline-Cole, R. A. and Abdulkadir, A. (eds) *Perspectives on Land Administration and Development in Northern Nigeria*, Kano, Department of Geography, Bayero University, Kano.

Barbier, E. B. (2003) 'Upstream dams and downstream water allocation: the case of the Hadejia-Jama'are floodplain, northern Nigeria', *Water Resources Research*, vol. 39, no. 11, pp. 1311–1320.

Binns, T. (1995) 'Introduction', in Binns, T. (ed.) *People and Environment in Africa*, John Wiley & Sons, Chichester.

Bird, A. and Tanko, A. I. (2004) *Remote Sensing Report*, Technical Report submitted to DFID/JEWEL Project, Nigeria, July 2004.

Blench, R. (2004) *Natural Resources Conflicts in North-Central Nigeria: A Handbook and Case Studies*, Mallam Dendo, Cambridge.

Bullock, A. and Acreman, M. (2003) 'The role of wetlands in the hydrological cycle', *Hydrology and Earth System Sciences*, vol. 7, no. 3, pp. 358–389.

CADP (2007) *Environmental and Social Management Framework (ESMF) for the Commercial Agriculture Development Project (CADP) – Final Report*, Federal Ministry of Agriculture and Natural Resources, Abuja.

Erwin, K. L. (2009) 'Wetlands and global climate change: the role of wetland restoration in a changing world', *Wetlands and Ecological Management*, vol. 17, no. 1, pp. 471–484.

FAO (1997) *Irrigation Potential in Africa: A Basic Approach*, Land and Water Bulletin 4, FAO, Rome.

HJKYBTF, NHI and IUCN (2007) *Concept Paper: Reoptimization of the Tiga and Challawa Gorge Dams to Restore Livelihoods and Ecosystems in the Hadejia-Jama'are-KYB-Lake Chad Basin*, unpublished report.

Hollis, G. E., Pension, S. J., Thompson, J. R. and Sule, A. R. (1993a) 'Hydrology of the river basin', in Hollis, G. E., Adams, W. M. and Aminu-Kano, M. (eds) *The Hadejia-Nguru Wetlands: Environment, Economy and Sustainable Development of a Sahelian Floodplain Wetland*, IUCN, Gland.

Hollis, G. E., Adams, W. M. and Aminu Kano, M. (1993b) *The Hadejia-Nguru Wetlands: Environment, Economy and Sustainable Development of a Sahelian Floodplain Wetland*, IUCN, Gland.

IUCN/HNWCP (1999) *Water Management Options for the Hadejia-Jama'are-Yobe River Basin, Northern Nigeria*, unpublished report for Hadejia-Nguru Wetlands Conservation Project.

IUCN/NIWRMC (2011) *Catchment Management Plan for Integrated Water Resources Management and Water Efficiency of Lake Chad Basin in Nigeria*, report submitted to Komadugu Yobe Basin Project Phase 2, IUCN and NIWRMC, Abuja.

JEWEL (2003) *Poverty, Environment and Livelihoods Issues Relating to Common Pool Resources in the Hadejia-Nguru Wetlands: A report to the JEWEL project*, unpublished report.

JICA (1995) *The Study on the National Water Resources Master Plan, Sector Report*, vol. 2, part 1, Federal Ministry of Water Resources and Rural Development, Abuja.

Kimmage, K. (1991) 'The evolution of the "wheat trap": the Nigerian wheat boom', *Africa*, vol. 61, no. 4, pp. 471–501.

Kolawole, A., Scoones, I., Awogbade, M. O. and Voh, J. P. (1994) *Strategies for the Sustainable Use of Fadama Lands in Northern Nigeria*, Centre for Social and Economic Research, Ahmadu Bello University, Zaria and IIED, London.

Kwaghe, U. K., Tanko, A. I., Kida, M. I. and Chiroma, M. J. (2008) *Socioeconomic Baseline Report on Communities Selected for Intervention in Hadejia-Jama'are-Komadugu-Yobe Basin, North-Eastern, Nigeria*, unpublished baseline report submitted to the Hadejia-Jama'are-Komadugu-Yobe Basin Trust Fund (HJKYBTF).

Lenton, R. and Muller, M. (eds) (2009) *Integrated Water Resources Management in Practice: Better Water Management for Development*, Earthscan, London.

McCartney, M. (2009) 'Living with dams: managing the environmental impacts', *Water Policy*, vol. 11, no. 1, pp. 121–139.

McCartney, M. and King, J. (2011) *Use of Decision Support Systems to Improve Dam Planning and Dam Operation in Africa*, International Water Management Institute, Colombo, Sri Lanka.

Moore, D., Dore, J. and Gyawali, D. (2010) 'The World Commission on Dams 10+; revisiting the large dam controversy', *Water Alternatives*, vol. 3, no. 2, pp. 3–13.

Mortimore, M. J. (1993) 'The intensification of peri-urban agriculture: the Kano close settled zone 1964–1986', in Turner, B. E., Turner, R. W. and Hyden, G. (eds) *Population Growth and Agricultural Change in Africa*, University Press of Florida, Gainsville.

Mortimore, M. J., Essiet, E. U. and Patrick, S. (1990) *The Nature, Rate and Effective Limits of Intensification in the Small-Holder Farming System of the Kano Close-Settled Zone*, Federal Agricultural Coordinating Unit, Nigeria.

Newson, M. (1997) *Land, Water and Development: Sustainable Management of River Basin Systems*, Routledge, London and New York.

NHI/IUCN (2007) Concept Paper: Reoptimization of Tiga and Challawa Gorge Dams to Restore Human Livelihoods and Ecosystems in the Hadejia-Jama'are-KYB-Lake Chad Basin: A Component of the Global Investigation of Techniques to Reoptimize Major Water Management Systems to Restore Ecosystems and Livelihoods.

Olofin, E. A. (1987) *Some Aspects of the Physical Geography of the Kano Region and Related Human Responses*, Departmental Lecture Note Series 1, Geography Department, Bayero University, Kano.

Olofin, E. A. (1991) 'Dam-induced changes in sediment characteristics of two savanna channels', in Olofin, E. A. and Patrick, S. (eds) *Land Administration and Development in Northern Nigeria: Case Studies*, Department of Geography, Bayero University, Kano.

Olofin, E. A. (1994) 'Dam construction and fadama development', in Kolawole, A., Scoones, I., Awogbade, M. O. and Voh, J. P. (eds) *Strategies for the Sustainable Use of Fadama Lands in Northern Nigeria*, Centre for Social and Economic Research, Ahmadu Bello University, Zaria and International Institute for Environment and Development, London.

Olomoda, I. A. (2003) *The Impact of Hydrological Information Services on Integrated Water Resources Management and Development: Niger River Basin Case Study*, Sixth Water Information Summit: Breaking the barriers, let water information flow!, Waterweb Consortium and IRC International Water and Sanitation Centre, 9–12 September, Delft, the Netherlands.

Olukesusi, F. (2011) *Socio-economic Importance of Dryland Management in West Africa*, unpublished paper presented at an international workshop on the development of proposals for the establishment of Centre for Dryland Agriculture (CDA), Bayero University, Kano.

Oyebande, L. (2003) *Appraisal of Hydrological Information on the Hadejia Jama'are Komadugu Yobe Basin in Light of JEWEL Project Needs*, a report to the JEWEL Project on the outcome of a two-week independent appraisal in the HJKYB between 20 May and 10 June 2003.

Polet, G. and Thompson, J. R. (1996) 'Maintaining the floods: hydrological and institutional aspects of managing the Komadugu-Yobe River Basin and its floodplain wetlands', in Acreman, M. C. and Hollis, G. E. (eds) *Hydrological Management and Wetland Conservation in Sub-Saharan Africa*, IUCN, Gland.

Ramsar (2010) *River Basin Management: Integrating Wetland Conservation and Wise Use into River Basin Management*, Ramsar handbooks for the wise use of wetlands, 4th edition, vol. 9, Ramsar Convention Secretariat, Gland.

Ramsar (2012) The Montreux record, http://www.ramsar.org/cda/en/ramsar-documents-montreux/main/ramsar/1-31-118_4000_0__ (accessed 30 October 2012).

Scoones, I., Reij, C. and Toulmin, C. (1996) 'Sustaining the soil', in Reij, C., Scoones, I. and Toulmin, C. (eds) *Sustaining the Soil: Indigenous Soil and Water Conservation in Africa*, Earthscan, London.

Tanko, A. I. (1996) *Improving Land and Water Resources Management in the Komadugu Yobe River Basin – North Eastern Nigeria and South Eastern Niger (Phase 1: Improving the Institutional Framework for Water Management in the Komadugu Yobe Basin)*, a technical mid-term report submitted to the FMWR-IUCN-NCF Komadugu Yobe Basin Project.

Tanko, A. I. (1999) *Changes in Soil and Water Quality, and Implications for Sustainable Irrigation in the Kano River Project, Kano State, Nigeria*, unpublished PhD thesis, Geography Department, Bayero University, Kano.

Tanko, A. I. (2010) 'Mega dams for irrigation in Nigeria: nature, dimension and geographies of impact', in Brunn, S. (ed.) *Mega Engineering Projects in the World*, Springer, Dordrecht.

Tanko, A. I., Kwaghe, U. K., Kida, M. I. and Chiroma, M. J. (2010) *Socioeconomic Impact Assessment Report on Communities that Benefitted from Intervention in Hadejia-Jama'are-Komadugu-Yobe Basin, North-eastern, Nigeria*, unpublished report submitted to the Hadejia-Jama'are-Komadugu-Yobe Basin Trust Fund (HJKYBTF).

Tarhule, A. and Woo, M. (1997) 'Characteristics and use of shallow wells in a stream fadama: a case study in northern Nigeria', *Applied Geomorphology*, vol. 17, no. 1, pp. 429–442.

Thomas, D. (1996) 'Water management and rural development in the Hadejia-Nguru Wetlands, North-East Nigeria', in Acreman, M. C. and Hollis, G. E. (eds) *Hydrological Management and Wetland Conservation in Sub-Saharan Africa*, IUCN, Gland.

UNDP (1998) *Nigerian Human Development Report, 1998*, UNDP, Lagos.

USAID (2007) *Preliminary Livelihoods Zoning: Northern Nigeria, A Special Report by the Famine Early Warning Systems Network*, unpublished report prepared by Chemonics for Review by the USAID.

WCD (2000) *Dams and Development*, Earthscan, London.

World Bank (2003) *Second National Fadama Development Project: Project Appraisal Document*, World Bank, Washington, DC.

10   Wetlands and rice development in West Africa

Paul Kiepe and Jonne Rodenburg

Summary

Inland valleys, which are seasonally to permanently flooded stream valleys, provide a range of ecosystem services (ESS). These landscapes play a strategic role in the regional efforts to attain food security and alleviate poverty. Besides agricultural production (mainly rice), inland valleys are also important suppliers of water, forest, forage, hunting and fishing resources to local communities and these ecosystems often harbour a high degree of plant and animal biodiversity, worthy of conservation. Only a fraction of the total inland valley area in West Africa should suffice to provide enough rice to feed the entire region and the remaining part could then be managed with other ESS as a priority, thereby addressing the needs for biodiversity conservation, protection against downstream flooding, natural buffers for excess water or for other livelihoods like pastoralism. This chapter describes approaches to achieve this, based on 40 years of experience of AfricaRice and partners, in particular through the work carried out under the umbrella of the Inland Valley Consortium (IVC) (currently Inland Valley Community of Practice). It is essential to use a systematic analysis approach for the selection of 'best-bet' inland valleys for rice production and to implement conservation regulations, and monitoring and evaluation mechanisms, for the protection of those valleys that are either too vulnerable or too valuable for agricultural development. Selection criteria for such 'best-bet' production valleys should include economic, social and biophysical characteristics and the selection of valleys should be broadly supported by the local communities depending on them. The same approach could be used to identify locations within the valley that could be used for crop production and those that should continue to fulfil other ecosystem functions, so that a specific valley has a multifunctional pattern of land use which will help ensure sustainability.

Introduction

Inland valleys are seasonally to permanently flooded stream valleys ranging in size from just a few hectares to several hundred. They constitute important agricultural and hydrological assets that provide a range of ESS, notably in hydrological regulation but also in terms of a range of livelihood benefits.

With careful interventions, the inland valleys could contribute much more to the national objectives of attaining food security and poverty alleviation in West Africa than they do today. In the Guinea and Sudan Savannah and the Equatorial Forest zones of West Africa rice is grown in the valley bottoms during the wet season and land is often left fallow during the dry season (Figure 10.1). If the rainy season is long enough (>5 months) or if residual soil moisture in the valley bottom is sufficient, farmers use the dry season to grow groundnuts, soybeans, maize, potatoes or vegetables, while in improved lowlands, where water can be controlled through drainage and storage facilities enabling off-season irrigation, rice is often followed by a second rice crop or vegetables (e.g. Windmeijer and Andriesse, 1993; Kent *et al.*, 2001). Besides rice production, inland valleys are important to local communities for the supply of a variety of other goods, such as forest, forage, hunting and fishing resources (e.g. Adams, 1993).

The Consortium for the Sustainable Use of Inland Valley Agro-Ecosystems in Sub-Saharan Africa (in short the IVC), convened by the Africa Rice Center (AfricaRice, formerly WARDA – the West African Rice Development Association) is composed of ten West African national agricultural research and extension systems (NARES) and a number of international (International Institute of Tropical Agriculture (IITA), International Livestock Research Institute (ILRI), the International Water Management Institute (IWMI), the Food and Agriculture Organization of the United Nations (FAO) and West and Central African Council for Agricultural Research and Development (CORAF)) and advanced research institutes (Centre de Coopération Internationale en Recherche Agronomique (CIRAD), Wageningen University and Research Centre). It was founded in 1993 with the objective of developing, through concerted and coordinated action, technologies and operational support systems for intensified but sustainable use of inland valleys in sub-Saharan Africa (SSA). The IVC uses a multidisciplinary scientific approach which aims to:

- determine the agro-ecosystem potential of inland valleys based on an integrated characterization and classification;
- identify means to achieve this potential by targeting research activities, by developing technological innovations and by transferring them to the plots, landscapes and watersheds; and
- build on already available resources such as inland valley ecosystems, inland valley developments, and local knowledge and innovations.

Recent IVC projects have focused on rehabilitation of abandoned or sub-optimally functioning inland valley developments, generating technologies based on local knowledge and innovations, and participatory valuation of inland valley ESS. They have sought to achieve sustainable productivity improvements for rice, by targeting water and weed management, exploring possibilities for the integration of rice–fish and rice–vegetable production. Overall, the goal of this work is to enhance the productivity

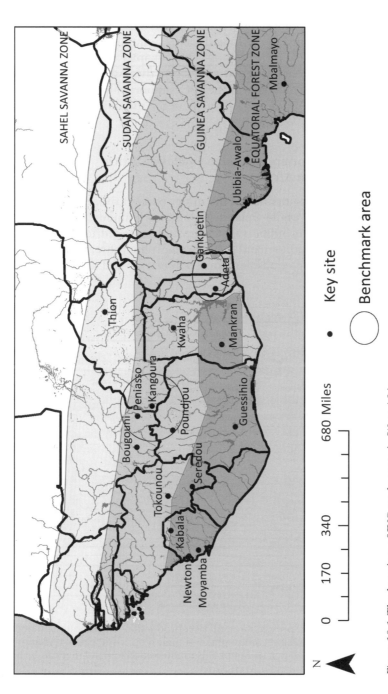

SAHEL SAVANNA ZONE

SUDAN SAVANNA ZONE

GUINEA SAVANNA ZONE

EQUATORIAL FOREST ZONE

Thion

Bougouni
Peniasso
Kangoura
Poundjou
Tokounou
Seredou
Kabala
Newton
Moyamba
Guessihio
Mankran
Kwaha
Gankpetin
Adeta
Ubibia-Awalo
Mbalmayo

• Key site

◯ Benchmark area

N

0   170   340   680 Miles

Figure 10.1 The location of IVC study areas in West Africa

and competitiveness of inland valley lowlands through sustainable intensification and diversification or agricultural productivity and product value
chain development, while conserving land and water resources. Since 2011,
the IVC has continued as the Inland Valley Community of Practice, which
better reflects its new *modus operandi*. This chapter describes some of the
important insights that have been gained over the past two decades through
work carried out by members of the IVC, supported by studies from other
parts of the world, in order to help achieve the sustainable management of
the inland valleys in West Africa.

The growing importance of inland valleys for rice cultivation in West Africa

Inland valleys can be defined as the upper parts of river drainage systems,
and comprise the whole upland–lowland continuum (Windmeijer and
Andriesse, 1993), including rain-fed uplands, rain-fed to intensified lowlands in the valley bottom and the hydromorphic (phreatic) fringes, comprising the (sloping) transition zone between upland and lowland areas
(Figure 10.2).

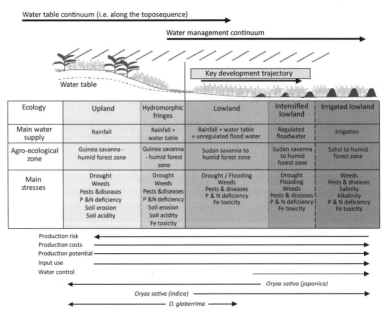

Figure 10.2 Schematic presentation of rice production environments along the
upland–lowland continuum, and their biophysical characterization: the
ecological range from uplands to intensified lowland that can be found
in inland valleys in West Africa

Sources: Andriesse *et al.*, 1994; Thiombiano *et al.*, 1996; Kiepe, 2006; and Wopereis and
Defoer, 2007

The morphology of inland valleys can vary as a result of climate, geology and geomorphology. There are many different cross-section shapes of inland valleys, but probably the most frequently observed morphology types are (Figure 10.3):

- rectilinear, broad valleys with gentle and straight slopes (<3 per cent);
- concave, relatively narrow valleys with concave side slopes (3–8 per cent); and
- convex, with moderately steep convex side slopes and flat narrow (20–400 m) valley bottoms.

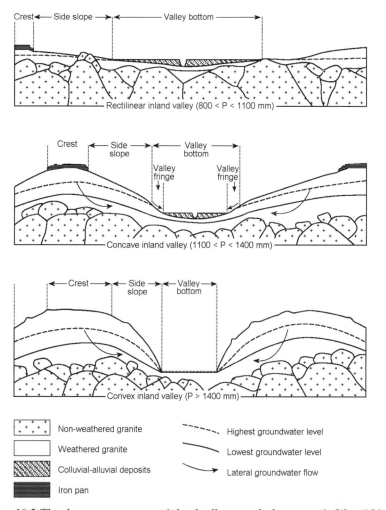

Figure 10.3 The three most common inland valley morphology types in West Africa, developed on granite-gneiss complexes under different rainfall regimes

Source: Raunet, 1985

Morphology, hydrology and climate together determine the depth, duration and frequency of flooding of the valley bottom, which in turn determine the suitability of the valley bottom for rice production. These physical conditions also determine which water management systems are likely to be most effective. There are five main systems:

- traditional random-basin system;
- central-drain system;
- interceptor-canal system;
- head-bund system;
- contour-bund system.
  (Oosterbaan *et al.*, 1987; Figure 10.4)

The traditional random-basin system comprises a partitioning into rectangular plots, surrounded by small bunds, where the water level within the plots is regulated by opening up the bunds. The central-drain system (Figure 10.4A) attempts to improve the traditional random-basin system by using a drainage canal to drain the valley-bottom. The interceptor-canal system (Figure 10.4B) comprises two interception canals in the fringes that run parallel to the central stream. The central stream can feed these interceptor canals through contour drains and the rice can be flooded from the interceptor canals while these canals also protect the rice fields from uncontrollable floods or runoff. The head-bund system (Figure 10.4C) has head bunds built perpendicular to the stream to create water reservoirs from which the rice can be flooded, using contour canals. The contour-bund system (Figure 10.4D) divides the valley bottom by contour bunds positioned transversely to the stream. Each contour bund has an outlet enabling water to flow from one field to another to flood the rice (Windmeijer and Andriesse, 1993).

The total inland valley area of tropical Africa is estimated to be 130 million ha (AfricaRice, 2008). The estimated total area of inland valleys in West Africa ranges from 38 to 84 million ha, or 12–27 per cent of the total area (Windmeijer and Andriesse, 1993) of which the valley bottoms and their hydromorphic fringes comprise approximately 22 to 52 million ha (Andriesse *et al.*, 1994). Inland valleys are abundantly available and common landscapes in West Africa that generally provide a multitude of ESS. Environments like these, in particular the valley bottoms (also known as *bas-fonds*, fadamas, inland swamps), generally have a high agricultural production potential due to more reliable water availability compared to rain-fed uplands (Andriesse *et al.*, 1994). However, the only major crop that can be grown under these temporary flooded conditions during the rainy season is rice (e.g. Andriesse and Fresco, 1991). Rice is a versatile crop that can be grown along the upland–lowland continuum, from the freely drained soils of the upland crests to the flooded conditions found in the valley bottoms, depending on the species (*O. sativa* or *O. glaberrima*) and sub-species (*japonica* or *indica*) (Figure 10.2). Rice is particularly suited to unregulated wetlands, where the environment can be flooded

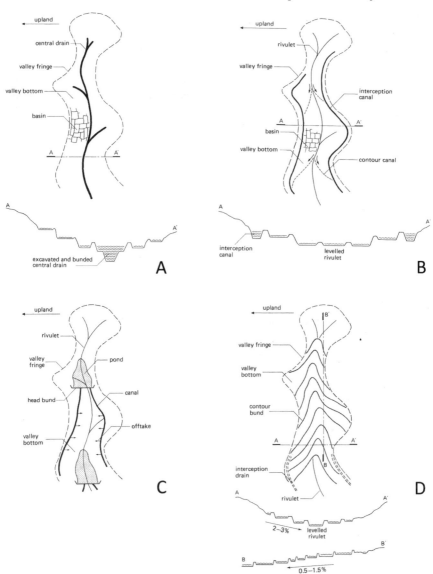

Figure 10.4 Water management systems of improved inland valleys: the central-drain system (A); the interceptor-canal system (B); the head-bund system (C); and the contour-bund system (D)

Source: Oosterbaan *et al.*, 1987; copied from Windmeijer and Andriesse, 1993

and then dried in rapid succession. Inland valleys are therefore of particular strategic importance for the development of the rice sector in Africa (e.g. Sakurai, 2006; Balasubramanian *et al.*, 2007), even more so because these landscapes constitute such a huge and yet largely unexploited area. Rice production in SSA (total production excluding northern African countries such as Egypt and Sudan is 18.45 million tonnes) has increased by 170 per cent over the past 30 years (FAO, 2010). This is partly due to an increase of 105 per cent in the area under rice production (total for SSA: 8.58 million ha; FAO, 2010) in the same period and partly because of a productivity increase following the development, introduction and implementation of improved production strategies, such as integrated crop management (e.g. Wopereis and Defoer, 2007) and new technologies, such as adapted and high yielding cultivars (e.g. Dalton and Guei, 2003). New Rice for Africa (NERICA) cultivars adapted to inland valley wetlands, for instance, have recently been developed (Sié *et al.*, 2008) and released in Burkina Faso, Mali, the Gambia (e.g. Balasubramanian *et al.*, 2007), Benin, Cameroon, Guinea, Liberia, Niger, Nigeria, Chad, Togo and Sierra Leone (M. Sié, *personal communication*).

Traditionally, inland valley wetlands in West Africa have been used for the collection of water and wild plant and animal resources, and have been little exploited for agricultural production (Adams, 1993). Reasons for this are plentiful: inland valley bottoms are difficult to manage and often harbour many water-borne human diseases such as bilharzia and malaria (Windmeijer and Andriesse, 1993). Moreover, inland valley exploitation is often hampered by complex or unfavourable (short-term) land tenure arrangements (e.g. Fu *et al.*, 2010; Oladele *et al.*, 2010) or customary beliefs that prohibit the use of such ecosystems. However, the last two decades have seen a trend towards increased inland valley use. Spontaneous exploration of inland valleys by subsistence farmers has been driven by declining fertility of upland soils due to reduced fallowing and unsustainable farming practices such as shifting cultivation (Windmeijer and Andriesse, 1993). Inland valleys have, in general, higher water availability and higher inherent soil fertility than the degraded upland soils, although soil fertility should still be considered poor compared to high input systems. While average upland rice yields in SSA stagnate at around 1 tonnes/ha (e.g. Rodenburg and Demont, 2009), it has been optimistically estimated that with good management, inland valley rice can produce up to ten times such yields (Wakatsuki and Masunaga, 2005). While 10 tonnes/ha will be difficult to obtain, improvements to attain at least five to six times the current average yields would be a realistic aim. Such improvements are centred on water, weed and soil fertility management. Relatively simple management practices, such as bunding and land levelling, significantly improve rice yields in inland valleys, not only by improving water management, but also through improved efficiency of fertilizer use and suppression of weeds (Becker and Johnson, 2001; Toure *et al.*, 2009) that are not adapted to permanently flooded conditions (e.g. Kent and Johnson, 2001). Further yield improvements can be expected from the introduction of improved, weed-competitive and adapted rice cultivars (e.g. Saito *et al.*, 2010). For instance, some NERICA cultivars adapted to lowland conditions are more weed competitive (Rodenburg *et al.*, 2009) and have a high yield potential (Sié *et al.*, 2008) and are, therefore, good candidate cultivars to raise productivity in these environments. The next generation of improved rice cultivars generated by AfricaRice and partners are named Advanced Rice for Africa (ARICA) and will be available for distribution in the next two or three years.

Global changes have given a new impetus to inland valley exploitation. Although there are a number of conflicting projections, in particular for the Sahel (e.g. Cook and Vizy 2006; Hoerling *et al.*, 2006; Biasutti *et al.*, 2008), climate trends suggest variability in rainfall in Africa will increase with drier conditions in the Sahel (Giannini *et al.*, 2008) and wetter conditions in equatorial zones (Christensen *et al.*, 2007). As the spatial distribution of future rainfall is highly uncertain (Giannini *et al.*, 2008), particularly for the Sahel, the value of a secure harvest from a wetland-produced crop becomes increasingly important for many countries in this region. Inland valley wetlands naturally have a higher availability of water, which in a progressively drier agricultural environment reduces the risk of crop failure. However, climate change also poses an ecological and hydrological threat for wetlands as they are sensitive to changes in water supply (both quantity and quality), while the prevention of wetland degradation requires strategies of adaptive management (Erwin, 2009).

In the past decade the market share for rice has increased in West Africa due to the growing local and regional demand for this staple food and declining availability of global stocks for export, leading to increasing farm-gate prices from an estimated US$255/ton in 1999 to US$445/ton in 2009 (FAO, 2010). This price increase encouraged many small-scale farmers to take up rice production. Between 1999 and 2009, sub-regional rice imports, mainly from Asia, decreased from an estimated 2.5 to 2.1 million tonnes but estimated production in West Africa increased from 7.1 to 10.3 million tonnes. At the same time, continuous migration to urban areas, in particular by young men, reduces labour availability in the rural areas making it difficult to sustain the expansion of rice production. Increasing productivity is the only way out of this situation and inland valley wetlands offer a potential avenue for this.

Sustainable productivity increase in inland valley rice production systems

Inland valley development

In order to exploit the potential of inland valleys to achieve the necessary increase in production and productivity, these environments should be brought into cultivation, bearing in mind the overall need for sustainability and the avoidance of degradation. Development of these valleys, as presently practised with support from IVC members, involves a close interaction by project actors, government or donor agencies, private sector contractors and the local communities. The emphasis now is much more on building community ownership of the valley development than was the case in the past when local government agencies just bulldozed areas without engaging with the communities in advance. Now the process starts with a discussion amongst the farmers and other key stakeholders about whether inland valley development is wanted and if so where this should be. A key consideration is the issue of the land rights of interested farmers in the valley sites proposed for development. A multi-criteria process is followed in order to identify the development sites with consideration given to a minimum size of the inland valley to make the development profitable, the presence of a water course, the availability of stones for bund construction to avoid running into high construction costs and secure land tenure to avoid future legal battles.

To empower farmers and facilitate discussions about development steps, farmers are organized into groups – which in the long term become water management/water user groups responsible for the management of irrigation water and drainage. These groups, in conjunction with the agency supporting the development, agree a development plan in terms of land development tasks, such as clearance of vegetation, construction of dams or other water management structures, as well as agreeing the development with downstream communities. The farmer groups also agree work responsibilities amongst their members which can involve the collection of stone and other materials for the work, as well as following up on the bulldozer work and the construction of channels for water management. Plots are allocated by the farmer groups to their members based on the work individuals have performed.

In all the necessary steps, from land clearing to development and actual crop production, a number of constraints are frequently encountered (Box 10.1). In the early stages, a lack of infrastructure (e.g. markets and roads), water management technologies and labour (to clear vegetation), water-borne human diseases (e.g. bilharzia and malaria) and high weed pressure are the general constraints observed (Andriesse *et al.*, 1994). After clearance and land development, a set of production constraints also need to be overcome in order to benefit from the inherent inland valley potential. The current actual (estimated) yields obtained from inland valleys across Africa (1.4 tonnes/ha according to Rodenburg and Demont, 2009) are far from the attainable yield of inland valley production systems (>10 tonnes/ha). From a survey conducted in eight West African countries, the top ten production constraints comprise both biophysical (e.g. unfavourable soil structure, high weed pressure, lack of water control) and socio-economic (e.g. land tenure and lack of inputs, labour and credit) constraints (Thiombiano *et al.*, 1996). (This chapter focuses on the biophysical constraints as the socio-economic ones, although recognized as also important, are not yet included in the ongoing work.) Once control over water has been secured, key factors to raise productivity in inland valleys are weed management and soil fertility (Wopereis and Defoer, 2007) and to a lesser extent control of pests and diseases (Figure 10.2).

Box 10.1 Inland valley rice development and the socio-economic characteristics of farmers in the Adeta watershed, Togo

Source: courtesy of K. Egué, ITRA, Togo

The Adeta watershed is located in the Plateaux region in South-West Togo, 150 km west of Lome. Inland valleys in this region are exclusively used for crop production, essentially rice, for which the cropping calendar is scheduled according to rainfall patterns. Valleys of the Adeta watershed are typical semi-urban inland valleys, near the town of Adeta and adjacent to a busy road and surrounded by several villages. There is good market access for the agricultural products and transportation costs are relatively low. The watershed comprises two inland valleys, one of 81.5 ha and another of 23.4 ha. Mean annual rainfall is 1,400 mm. Land use can be characterized as 'traditional' and tools are rudimentary (e.g. hand-hoes and oxen ploughs). The rainfall regime and hydrology of the valleys favour production diversification because of the plentiful water supply in the dry season.

More than half of the farmers (54 per cent) are female and 71 per cent of the farmers are literate (primary school, 37 per cent; secondary school, 27 per cent; and functional literacy, 7 per cent). The economic activities are dominated by crop production (89 per cent), primarily rice. Secondary activities are livestock production and handicrafts for men, and trade and handicrafts for women. Less than 25 per cent of farmers benefit from guidance by the agricultural development services. Access to land is essentially through renting (48 per cent) and borrowing (40 per cent).

The high cost of seasonally constructed levees, which in some cases can reach up to one-third of the production costs, hampers the full use of inland valleys. The surface area per producer is relatively small (mean: 0.30 ha per person) while 94 per cent of farmers still call for external labour due to the high work load. Men generally carry out land clearing, tillage and threshing, whereas women specifically take care of sowing, weeding, transportation and winnowing.

Inputs, more or less regularly used by farmers, are as follows: 75 per cent use improved seed; 65 per cent use chemical fertilizers (nitrogen–phosphorus—potassium) and 21 per cent use pesticides (primarily herbicides). The lack of water control is the most critical constraint. Farmers use nine improved and five local rice cultivars. There are three types of rice establishment methods: transplanting, direct sowing to hills and broadcast sowing. Direct sowing, which involves drilling rice seeds in a specific arrangement such as lines, is the most common method because of the lower time requirements compared to transplanting. Rice yields range from 0.5 to 3.5 tonnes/ha while total production costs range from US$100 to US$300 per ha. With a local rice price of around US$1 per kg and an average field size of 0.3 ha, this leaves farmers with low net returns. (Assuming a strong correlation between production costs and yield, returns would range from US$120 to US$960 per cropping season.) Farmers do not have access to agricultural credit systems and much irrigation infrastructure (e.g. inlets, outlets) is abandoned after project withdrawal due to their high cost of maintenance, a lack of involvement of producers in the design and implementation of inland valley development, and the high land rent levies.

Challenges

By far the biggest constraint to rice cultivation in inland valleys is competition from weeds. The weed flora in this ecology is dominated by grasses and sedges (Johnson, 1997; Wopereis *et al.*, 2007; Rodenburg and Johnson, 2009). Another, emerging, problem weed in inland valleys is the facultative parasitic plant *Rhamphicarpa fistulosa* (Rodenburg *et al.*, 2010) which uses parasitism through underground root-to-root connections (Ouédraogo *et al.*, 1999). Parasitic infection reduces the reproductive ability of the host plant (e.g. rice) by 63 per cent on average; farmers often are forced to abandon their fields due to this (Rodenburg *et al.*, 2011).

A second, overarching, production constraint in inland valleys is the general low soil fertility. One important motive for farmers to descend from the relatively easy cultivatable uplands to the more labour intensive valley bottoms is the higher soil fertility status of the lowlands. This is mainly a consequence of soil deposition (in particular of silt and fine clay factions) by erosion from the adjacent uplands (Ogban and Babalola, 2003). These processes have, for instance, resulted in higher levels of exchangeable bases (Kyuma, 1985), calcium and magnesium (Fagbami *et al.*, 1985) and phosphorus (Ogban and Babalola, 2003) compared to nearby uplands. However, in most inland valleys soil fertility is far from optimal for crop production. Although the soil fertility status varies considerably among agro-ecological zones (Issaka *et al.*, 1997), extensive soil sampling in inland valleys across West Africa has shown that a number of important soil fertility parameters (e.g. pH, cation exchange capacity (CEC), total carbon and nitrogen, available phosphorus) are low to very low (Issaka *et al.*, 1996), while micro-nutrients such as sulphur and zinc are also generally deficient (Buri *et al.*, 2000). Besides the low nutrient status, clay mineralogy in West African inland valleys is also poor (Abe *et al.*, 2006).

Another common soil fertility-related constraint to rice production in inland valleys in West Africa is iron toxicity (Becker and Asch, 2005; Audebert and Fofana, 2009). Iron toxicity is the common term for a complex nutrient disorder expressed when excessive iron in the soil solution is combined with the specific water-logged conditions that are typical for poorly drained inland valleys in West Africa (Narteh and Sahrawat, 1999). Under such circumstances the plant takes up more soluble iron (Fe(II)+) than is needed, leading to iron accumulation in the leaves beyond a critical level. This results in leaves turning reddish-brown or yellow (leaf 'bronzing') and eventually dying off, which in turn negatively impacts crop yield (e.g. Becker and Asch, 2005). Yield reductions caused by iron toxicity in lowland rice in West Africa typically are in the order of 40–45 per cent, depending on the severity of the problem, rice cultivar tolerance, water and soil management and the availability of other soil nutrients (Audebert and Fofana, 2009).

Challenges addressed

Through improved integrated water, weed and soil fertility management in inland valley rice production systems, yields can increase considerably. Bunding, for instance, improves water and weed management and increases nutrient use efficiencies, in particular in relatively well-drained inland valleys. It has been shown that this relatively simple technology could increase rice yields by 40 per cent and reduce weed pressure by 25 per cent across environments (Becker and Johnson, 2001; Toure *et al.*, 2009). A multi-site study also showed that application of mineral nitrogen further increased yields by 18 per cent in bunded fields (Becker and Johnson, 2001). More work carried out by AfricaRice has shown that for rice productivity in the inland valleys to increase, a flexible, locally adapted and integrated approach is needed (Wopereis and Defoer, 2007). AfricaRice developed a curriculum for Participatory Learning and Action Research (PLAR) for Integrated Rice Management (IRM) in inland valleys, which is essentially an approach to improve farmers' crop management, to stimulate farmer experimentation and to help identify researchable issues. This method consists of a technical manual (Wopereis *et al.*, 2007) and a facilitators' guide (Defoer *et al.*, 2004) containing modules on water, crop and pest management issues. Many PLAR-IRM modules have also been converted into farmer training videos (entitled 'Rice Advice') which have been translated into many local languages and widely distributed over the continent (AfricaRice, 2011).

Multifunctional character of inland valleys

Besides agricultural production, mainly rice and vegetables, as well as aquaculture and livestock production, inland valleys harbour a range of functions, delivering associated products and attributes (Adams, 1993). Important ESS of inland valleys are regulatory and support services comprising water storage, flood and erosion control, nutrient retention, micro-climate stabilization, and recreation and tourism. Besides food crops these environments deliver other provisioning services such as forest, wildlife and forage resources, fisheries, water, clay and sand, and crafts and construction materials, and they contribute to cultural services, notably biological diversity and local heritage (Dugan, 1990; Adams, 1993). The inland valley water resources and their biological diversity are probably the most important traditional services for local communities. A survey carried out in southern Benin and Togo showed that inland valleys are important locations for local communities to collect non-agricultural plant resources (Rodenburg et al., 2012). It was found that a large number of non-cultivated plant species are known by a variety of local community members (men and women, young and old) and collected for a wide range of purposes (e.g. food, medicine, construction, ceremonies). Many of these plant species have three or more uses (Table 10.1).

The multifunctional character of inland valleys renders them attractive for exploitation in a variety of ways. Indeed in recent decades the economic opportunities of these areas have been recognized and both large- and small-scale investments have been made to access and develop these areas. However, indiscriminate development of wetlands may lead to degradation of natural resources, which in turn jeopardizes their unique and plural ecosystem functions (e.g. Dixon and Wood, 2003). In particular, the trade-off between land use and conservation of natural resources is critical for the sustainable, multiple use of African wetlands (e.g. McCartney and Houghton-Carr, 2009). Inland valley developments therefore need to be planned carefully and implemented in close association with the local users. Unfortunately the importance of wetlands for local communities in Africa has often been overlooked in policy and planning (Silvius et al., 2000). Studying the local use and management of ecosystem functions could help raise awareness of the importance of inland valley systems and generate recommendations for sustainable use (e.g. Rodenburg et al., 2012).

Different ESS do not necessarily compete with one another. While local users have indicated agriculture as one of the most important causes for decline in plant biodiversity, agricultural fields were also considered as one of the most important locations for the collection of useful, non-cultivated plants (Rodenburg et al., 2012). This is partly because of the lack of remaining forests and partly because of the relatively easy access to agricultural fields. On top of that, the weed flora comprises many useful species. Some of the useful plants mentioned by local people are in fact known as weedy

Table 10.1 Plant species with three or more uses prioritized by local users (n=106) in six villages in Benin and Togo (2005–2006)

Species	Family	Local use
Adansonia digitata	Malvaceae	Fo, M, H
Anogeissus leiocarpa	Combretaceae	M, H, T, D
Antiaris toxicaria	Moraceae	Fi, C, H
Azadirachta indica	Meliaceae	M, Fi, C, H, T
Blighia sapida	Sapindaceae	Fo, M, Fi, C, T
Caesalpinia pulcherrima	Fabaceae	M, H, T (W)
Ceiba pentandra	Bombacaceae	C, H, FH
Citrus aurantifolia	Rutaceae	Fo, M, Fi,
Citrus aurantium	Rutaceae	Fo, M, Fi
Cocos nucifera	Arecaceae	Fo, M, Fi, C, H, T
Elaeis guineensis	Arecaceae	Fo, M, Fi, C, H, T, Cp, FH
Holarrhena floribunda	Apocynaceae	M, Fi, C, H, T
Khaya senegalensis	Meliaceae	Fo, M, Fu, T, D
Lonchocarpus sericeus	Fabaceae	M, Fi, H, Cp (W)
Mangifera indica	Anacardiaceae	Fo, M, Fi
Milicia excelsa	Moraceae	M, Fi, C, H, T, FH
Morinda lucida	Rubiaceae	M, Fi, H
Newbouldia laevis	Bignoniaceae	M, Fi, H, T
Parkia biglobosa	Fabaceae	Fo, M, Fi, T, C
Psidium guajava	Myrtaceae	M, Fo, Fi, H
Pterocarpus erinaceus	Fabaceae	M, Fi, C, H, T
Senna siamea	Fabaceae	M, Fi, C, H (W)
Strychnos innocua	Loganiaceae	M, Fi, T
Tectona grandis	Lamiaceae	M, Fi, C, H
Zanthoxylum zanthoxyloides	Rutaceae	Fo, Fi, C, Tr

Key: Fo = food, Fi = firewood, M = medicinal, C = construction, H = household, Cp = crop protection (including agroforestry, soil water conservation and pesticides), T = traditions and ceremonies, FH = fishing and hunting, (W) = known as a weed or invasive species.

Source: adapted from Rodenburg *et al.*, 2012

or invasive species. Subsistence farmers in West Africa often recognize these useful weed species during weeding and either keep them apart or leave them untouched (Rodenburg and Johnson, 2009). In addition, at field clearing, useful species (primarily woody species) are also maintained (Rodenburg *et al.*, 2012), as often observed in West Africa (e.g. Leach, 1991;

Madge, 1995; Kristensen and Lykke, 2003). This is an increasingly common strategy to deal with the decline of forests (Shepherd, 1992). In addition, Rodenburg *et al.* (2012) observed the establishment and management of a community garden where useful species (around 300) are nurtured. This was located in the village but included species that would previously have been found in undisturbed inland valleys. Development of such a garden saves local users the time, effort and uncertainty of collection, and it alleviates the pressure on the natural, relatively undisturbed, vegetation such as sacred forests. These observations support the idea that different ESS (agricultural production and biodiversity conservation, use and management) within inland valley landscapes may be exploited synergistically. However, it requires participation of local communities in any development or conservation initiative to align these multiple interests. Local communities are the primary users and managers of plant resources and therefore the most important stakeholders. Their knowledge and needs need to be acknowledged to establish a basis for sustainable land use (Rodenburg *et al.*, 2012). This synergy could be achieved by using multi-criteria decision making (MCDM) methods based on the insight that stakeholders use multiple criteria for decision making (e.g. Raj, 1995). McCartney and Houghton-Carr (2009) proposed the use of an index (Working Wetland Potential (WWP) index), which can be used to evaluate the potential of different agricultural activities in a given valley and to find synergies or compromises between different users and ESS. Local stakeholders should be involved in land use planning for inland valleys and this could be achieved by using the so-called multi-stakeholder platforms (MSPs) as demonstrated by Warner (2006).

Complex physical and social environment

The development of inland valleys for rice cultivation requires considerable flexibility as these areas are biophysically and socio-economically extremely diverse and complex (e.g. Andriesse *et al.*, 1994). The environmental diversity of inland valleys and their heterogeneous social context renders the development of inland valleys complex (Boxes 10.2 and 10.3).

Regarding the biophysical complexity, it can be stated that every valley is unique. So, there is no 'off-the-shelf' technology that can be applied to each valley. Regulation of the water regime – be it to control flooding, to optimize its use for irrigation, or to conserve water to be used later in the season – needs to be tailored to the local conditions. With regard to the social environment the situation may be even more complex. Land tenure arrangements in inland valleys differs from place to place and are often complex (e.g. Fu *et al.*, 2010) and impede investments. Ownership varies between and within countries. Valleys can be owned by families, individuals, villages, states and so on (Box 10.3). The farmers working in the inland valleys are often not the owners of the land and therefore not always the beneficiaries of investments targeted for inland valley

Box 10.2 Socio-economic and environmental constraints of inland valley land use in Benin

Source: courtesy of J. F. Djagba, C. Houssou and P. Kiepe (Université d'Abomey Calavi/AfricaRice)

The socio-economic and environmental constraints of inland valley land use are many and their links complex. To this end, 300 farmers from 30 inland valleys, associated to 30 villages in 12 districts of Benin, were interviewed. Twenty-one of the 30 inland valleys were 'improved'. Improved inland valleys have simple water storage facilities, bunds along the contours and a main irrigation canal from where bunded fields are provided with water using gravity irrigation. The unimproved inland valleys lack such water storage and irrigation canals with inlets and outlets. The occupation rate of the improved area in inland valleys was 90 per cent. The farm size in inland valleys was on average 0.4 ha without a significant difference between improved and unimproved inland valleys. Monoculture rice cropping is the dominant system (75 per cent of the cases) in both types of system. The average yield of rice is 2.0 tonnes/ha (max. 6.8 tonnes/ha), 1.3 tonnes/ha in unimproved inland valleys and 2.2 tonnes/ha in improved ones. Besides inland valley improvement, differences in yield are caused by the agro-ecological zones and the use of chemical inputs. The average net income per hectare per crop (season) is 223,415 F CFA (around US$450), ranging from 64,068 F CFA (US$128) in unimproved inland valleys to 271,912 F CFA (US$544) in improved ones. This study identified three main contributors to profitability of agricultural activities in inland valleys: 1) improved water management; 2) access to agricultural credit; and 3) the exclusive use of family labour. The remaining constraints identified for the sustainable exploitation of inland valleys are: 1) human health problems; 2) lack of agricultural equipment; 3) the lack of a functional value chain; 4) the occurrence of pests and weeds; 5) iron toxicity; and 6) land degradation and erosion.

Box 10.3 Land tenure arrangements in the inland valleys of Sierra Leone

Source: courtesy of M. Kandeh and A. S. Lamin, Land and Water Development Division, Ministry of Agriculture Forestry and Marine Resources, Sierra Leone

Land tenure arrangements in inland valleys were studied in 1996 in all seven regions of Sierra Leone using a sample size of 150 farmers per region (Table 10.2). Land tenure is a controversial issue in all the agricultural regions of Sierra Leone. The methods of acquiring agricultural land in the Western region (Freetown and suburbs) are quite different from those in the rural areas. In the rural areas, land is not purchased and owned as freehold. Instead, the major form of land tenure is family or community ownership. The traditional head, i.e. the chief or tribal leader, is the custodian of the land and empowered to oversee all transactions and settle land disputes. The percentage of farmers using family and community land ranges from 60 to 73 per cent. Second to communal ownership is acquisition of land by virtue of marriage or long-term settlement and cooperation with the local people (17–30 per cent). Land tenure sometimes poses serious problems when it comes to inland valley development especially when it involves improvement of water control. Inland valley wetlands are normally cultivated at several different locations upstream, midstream or downstream, depending on where settlements are located. In some cases it is impossible to implement development at any location without the collective consent of farmers in all the settlements along the valley. For example, in 1993, an EU-sponsored earth dam in an inland valley swamp in the Port Loko district was abandoned because of disagreement over the effects it would have on water availability for downstream farms. The problem was solved later, when all the villages were encouraged to form a farmers' organization sponsored by the project.

Table 10.2 Land tenure systems in inland valleys in Sierra Leone

Region	Land tenure systems			
	Family + community land	Leased or rented land	Freehold	Marriage or settlement
East	60	17	3	20
South	63	6	3	28
North	65	3	2	30
West	46	20	17	17
South-west	73	7	1	19
North-west	72	5	3	20
North-central	65	5	3	27

Source: survey data, May 1996; sample size 150 farmers/region

development. Land tenure arrangements can also affect gender relations. In most cases land is owned by men and cultivated by women, in particular when the land is of little value (e.g. rain-fed, low soil fertility, difficult to access). When the land is developed, and thereby its value increased, men can reclaim their rights. Inland valley development projects have to take these land tenure and gender relations into account when the aim is to benefit the poor and to treat men and women equally. The inland valley of Pegnasso, a village near Sikasso, southern Mali, was developed by the French Development Cooperation in 1994. The project responsible for the development of this inland valley deliberately chose to introduce only a partial development instead of complete development to avoid redistribution of the land among farmers. They reasoned that large investments in the valley would increase the value of the land too much and cause conflicts. Women, who owned or used the land before the development, risked being chased away when the value of their land increased, and would consequently not profit from the project. This partial development entailed the construction of a small (concrete) water-retention structure and one main water inlet and central irrigation canal from which water could be directed to the neighbouring farmers' fields with simple bunds for within-field water management. These relatively simple improvements enabled farmers to make better and more prolonged use of the available water for increased rice production, which in that area is mainly the responsibility of women. The project therefore succeeded in benefiting the community and strengthening the position of women.

Another example showing how women can benefit equally from inland valley development comes from the Blétou valley in South-west Burkina Faso. This inland valley, part of a larger system of interconnected inland valleys, was purely rain-fed and as a consequence lack of water retention was the main problem. Water was only available for crop production for a short time during and after a rainfall event, after which the water quickly drained to lower parts of the larger inland valley system. Consequently the main biophysical constraint to rice production was drought, and yields were erratic and low (often less than 1 ton/ha). In an IVC project, funded by the Common Fund for Commodities (CFC) and implemented between 2006 and 2009, AfricaRice and the Institut de l'Environnement et de Recherches Agricoles (INERA) proposed to improve water availability in that part of the inland valley that had the highest potential return. This was achieved by the installation of water retention structures consisting of bunds along the contour lines, covered with impermeable and resilient cloth and lateritic rocks. The project opted for a community–participatory development approach, whereby the beneficiaries would be those farmers that assisted in the construction of water-management structures. This reduced the costs and also provided the farmers with a sense of ownership. Since these structures mainly consisted of lateritic rocks, farmers were involved in collecting rocks. Women farmers comprised the main group (104 out of a total of 121) and they were rewarded in the partitioning of the plots when the inland valley development was completed. The plots were assigned by the whole group in a participatory manner as a function of each individual's contribution to the development work. This approach resulted in a fair distribution of the plots, doing justice to individual time investments and an optimal use of the developed part of the inland valley. Today, more than three years after completion, the same number of women continue to benefit from their investment.

Towards a sustainable inland valley development process

Sustainable use of inland valleys is important particularly in developing countries as it presumes to encompass both economic and ecological objectives, balancing local livelihoods with biodiversity conservation (e.g. Wyman and Stein, 2010). To attain development that balances different ESS and users, good communication between different stakeholders, at all levels, is an important prerequisite. Large investments in water management infrastructures for inland valley development in the 1970s have in most cases been abandoned because of the top-down approaches used for the selection, design and planning of these developments (e.g. Dries, 1991; Maconachie, 2008), or because traditional land-tenure arrangements were ignored (Brautigam, 1992) (Benin). However, the potential is considerable and Ahamidé et al. (2004) estimated that more than 87,000 ha of floodplains and inland valleys throughout Benin could be irrigated for agriculture. In the 1960s and 1970s many such irrigation schemes were developed by public investment corporations and development projects. For instance, the Benin-China Cooperation intervened in numerous medium-sized lowland areas (typically ranging from 25–150 ha) and a total of 1,400 ha were equipped with irrigation infrastructure (e.g. water collection structures, irrigation canals, drains, inlets and outlets, small bridges and cofferdams). Most of these projects failed and the irrigation schemes have been abandoned. One of the rare exceptions is the still functioning and extensively used Koussin-Lélé irrigation scheme, where farmers use 106 ha of developed land to grow rice with gravity irrigation. From a comparative study between this scheme and nearby schemes of Zonmon and Bamè (84 ha and 33 ha respectively) it was learned that careful selection of the valley, and the involvement and consultation of the main local stakeholders from the very start of the development effort (e.g. planning, design, implementation) to the actual use and management, were essential steps for success and sustainability (Djagba et al., 2012).

Based on the experiences of the IVC in the development of sustainable (agricultural) land use in the inland valley wetlands in West Africa, and sustainable rice production improvement, a step-wise approach is now advocated. This starts with the use of an inventory of inland valleys at a national scale with different levels of detail (Andriesse et al., 1994;

van Duivenbooden *et al.*, 1996). Geographical Information System (GIS) and remote sensing tools are useful for preliminary mapping and initial low resolution analyses (e.g. Thenkabail and Nolte, 1996; Gumma *et al.*, 2009; Chabi *et al.*, 2010). This can be followed by the proposed Integrated Transect Method (ITM) of van Duivenbooden *et al.* (1996) that uses surveys and farmer interviews conducted along transects across valleys to obtain information on the physical environment and land use forms. The 'best-bet' inland valleys should be selected because they score high on (agricultural) production potential and relatively low on other ESS such as biodiversity. Not all inland valleys are suitable for rice production. Such a thorough inventory using GIS/remote sensing down to the ITM, as well as careful planning, can avoid unnecessary destruction of wetlands and bad economic investments. Second, an environmental systems analysis should look into the upstream–downstream consequences of improving or developing inland valleys. Third, in areas that are selected, the development efforts should strive for active participation of all stakeholders (e.g. Perfecto and Vandermeer, 2008; Del Amo-Rodriguez *et al.*, 2010) in ways similar to the approaches proposed by McCartney and Houghton-Carr (2009) and Brockington (2007). The aim would be to take the social order into account, which means that all local stakeholders should play a role in the development process. In this step so-called MSPs could be created and community analysis of development options conducted. The next steps would include the participatory identification and mapping of hot spots for specific ESS within the upland–lowland continuum under consideration. This could lead to a locally developed, strategic, multiple-use plan for inland valley land use, with designated areas for agricultural production and areas that are maintained or managed to provide other ESS, including the conservation of biodiversity hotspots and sites of cultural importance. The result would be a putative, environmentally and socially sustainable pattern of land use, with spatially and temporally mixed land use forms similar to that proposed by Dixon and Wood (2003) for valley swamps in south-west Ethiopia. While there is some experience with the first steps of this approach (e.g. multi-scale characterizations using GIS and remote sensing tools), little published evidence yet exists to show that the other steps (e.g. environmental systems analysis and formation of MSPs) will definitely result in the desired sustainable use of inland valleys in West Africa.

Currently an IVC project called RAP (Realizing the Agricultural Potential of Inland Valley Lowlands in Sub-Saharan Africa) has started to work with the MSP approach in sites in Mali and Benin. This aims to build capacity and strengthen institutional change and development, and to analyse opportunities and risks related to agricultural production and environmental goods and services. Moreover, the project facilitates the development of innovative technologies for sustainable intensification and diversification of land and water resources, and builds competitive value chains

for rice and vegetable crops thereby enhancing market access for produc-
ers. The main target group are the resource-poor smallholder farmers and
the farmer organizations and local entrepreneurs, while the main service
providers are the 'change agents' from NARES to non-governmental organizations (NGOs). Preliminary results are promising, e.g. inland valleys are
now included in the socio-economic and development plan of a munici-
pality in one of the sites in Mali, and, in general, dynamic participatory
action research efforts have been initiated with the actors in the lowlands
based on the MSPs. These results need to be confirmed and their applicability should be tested at a larger scale before this approach is promoted as
a means to achieve sustainable use of inland valleys.

Once the inland valley has been selected, and the inland valley plan
has been made with the participation of all important local stakeholders,
the actual development, mainly aimed at improving control of water (e.g.
irrigation and drainage structures), can start. Due to the high diversity
of inland valleys, both biophysical and socio-economically, there is no
such thing as a 'blueprint' for inland valley development. The most suitable type of water-control structure for a specific inland valley depends
on various physical factors such as the size of the catchment area, soil
texture (determining hydrological behaviour) and the valley morphology as described before. Prior to the development of an inland valley, a
thorough diagnostic study should be carried out to determine the movement of water (in space and time) within the valley (Wopereis *et al.*, 2007).
DIARPA (Diagnostic Rapide de Pré-aménagement), a diagnostic interactive tool developed under the umbrella of the IVC (Legoupil *et al.*, 2001),
can be of use at this stage (Lidon *et al.*, 1998). This rapid hydraulic diagno-
sis of a lowland assesses the best type of intervention, at a certain location
and level of investment, to limit hydraulic risks and secure agricultural
production. Explanatory indicators for water dynamics (categorized under
pedological, topographic and hydrological indicators) have been selected,
characterized and validated for their use as an index when choosing types
of interventions (Table 10.3).

Conclusions: regional potentials, local approaches

Inland valleys are an integral part of the West African landscape. They comprise a huge and largely unexploited but also largely unprotected land area
with valuable water resources that could be used to boost rice production in
order to attain rice self-sufficiency and improve food security in the region.
However, inland valleys are highly diverse and complex systems, both biophysically and socio-economically, requiring locally adapted strategies and
technologies and stakeholder participatory approaches for their development and productivity enhancement. Apart from their crop production
potentials, mainly for rice and vegetables, inland valleys also provide a multitude of other ESS of local and regional relevance. While food security is

Table 10.3 DIARPA indicators and their threshold values for each type of intervention

Indicators	Type of interventions					
	Contour bunding	*Contour bunding with spill over*	*Water retention dykes without seepage barrier*	*Water retention dykes with seepage barrier*	*Diversion barrier to diffuse flow*	*Diversion barrier for the re-infiltration and restocking of the water table*
Pedological						
Permeability (m/s)	$<10^{-4}$	$<10^{-4}$	$<10^{-4}$	$<10^{-4}$	$<10^{-4}$	$<10^{-4}$
Impermeable layer depth	NI	NI	NI	<2m	NI	NI
Topographic						
Average longitudinal slope of the valley	<1%	<1%	<0.5%	<0.5%	<1%	<1%
Flow axis	No flow axis	NI	Noticeable flow axis	Noticeable flow axis	Narrow flow axis	Narrow flow axis
Hydrological						
Ten year maximum discharge, litres per second, per metre of valley bottom width	3×10^{-3} m^3 s^{-1}	25×10^{-3} m^3 s^{-1}	0.25–0.6 m^3 s^{-1}	0.25–0.6 m^3 s^{-1}	50×10^{-3} m^3 s^{-1}	50×10^{-3} m^3 s^{-1}
Depth of the groundwater at beginning of dry season	NI	NI	NI	NI	NI	<2m

Note: NI = not important

Source: courtesy of F. Blanchet and J. C. Legoupil

an important objective for the region, and inland valleys are of strategic importance to achieve this, care should be taken in the process of their selection and development. If only 10 per cent of all inland valleys in Africa were set aside for rice production, 2.2 million ha according to the most conservative estimation of the valley-bottom area, and average rice productivity raised to an attainable 6 tonnes/ha by using the IRM practices developed by AfricaRice and partners, this would produce over 13 million tonnes of rice, about the current sum of total rice production and consumption in West Africa. Hence, despite increasing consumption due to population growth and rising incomes, a fraction of the total inland valley area should suffice to provide enough rice to feed the entire sub-region. Consequently, a significant part (e.g. the remaining 90 per cent) could then be managed with other ESS as the priority. In this way the needs for biodiversity maintenance, wildlife sanctuaries, protection against downstream flooding, natural buffers for excess water or non-agricultural livelihoods could be met. With this mixed pattern of land use along stream valleys the prospects of ecological sustainability recommended by the Millennium Ecosystem Assessment (MA) would be enhanced.

Based on its 40 years of experience as Africa's centre of excellence in rice research and development, and nearly two decades of experience as the main coordinating institute of the IVC, the Africa Rice Center (AfricaRice) has developed a step-wise and bottom-up (participatory) approach for site selection, land use planning, design and implementation of rice production in inland valleys, as well as technologies and tools supporting this approach and farmer participatory methods to adapt technologies to local biophysical and socio-economic conditions. AfricaRice and partners have concluded that it is essential to use a systematic approach for the selection of 'best-bet' inland valleys for rice production and to implement conservation regulations and monitoring and evaluation mechanisms that will help protect those inland valleys that are either too vulnerable (e.g. due to erosion risks, loss of water resources or risks of social conflicts) or too valuable (in terms of other ESS such as biodiversity) for agricultural development. Selection criteria for such 'best-bet' production valleys should include economic, social and biophysical characteristics and the selection should be broadly supported by the local communities who depend on the inland valleys. Depending on the size, the same approach can be used to identify specific locations within the inland valley that could be used for crop production and those that should continue to fulfil other ecosystem functions, so that individual valleys have a multifunctional land use. For the sustainable development and use of the potentials offered by the inland valleys in West Africa, full participation of local stakeholders is required at all stages of decision making, development and implementation. This should lead to interventions that are flexible, locally adaptable, and culturally and socio-economically acceptable. In this way the supply of rice for West Africa can be assured for the future in a sustainable manner.

Acknowledgements

The authors gratefully acknowledge permission granted by Alterra (Wageningen UR) to reproduce copyrighted material from Windmeijer and Andriesse (1993) as Figure 10.4 in this chapter.

References

Abe, S. S., Masunaga, T., Yamamoto, S., Honna, T. and Wakatsuki, T. (2006) 'Comprehensive assessment of the clay mineralogical composition of lowland soils in West Africa', *Soil Science and Plant Nutrition*, vol. 52, iss. 4, pp. 479–488.

Adams, W. M. (1993) 'Indigenous use of wetlands and sustainable development in West-Africa', *Geographical Journal*, vol. 159, no. 2, pp. 209–218.

AfricaRice (2008) 'The West Africa Inland Valley Information System (WAIVIS)', www.africarice.org/WAIVIS (accessed 15 March 2010).

AfricaRice (2011) 'Rice Advice', www.africarice.org/warda/guide-video.asp (accessed 24 September 2012).

Ahamidé, B., Agbossou, E. and Igué, M. (2004) 'Recherche bibliographique sur la mise en valeur des bas-fonds au sud du Bénin', Département du Mono/Couffo, Ouémé/Plateau et Atlantique, Actes de l'atelier scientifique Sud et Centre, 11–12 Décembre, pp. 109–125.

Andriesse, W. and Fresco, L. O. (1991) 'A characterization of rice-growing environments in West Africa', *Agriculture, Ecosystems & Environment*, vol. 33, no. 4, pp. 377–395.

Andriesse, W., Fresco, L. O., Duivenbooden, N. and Windmeijer, P. N. (1994) 'Multi-scale characterization of inland valley agro-ecosystems in West Africa', *Netherlands Journal of Agricultural Science*, vol. 42, no. 2, pp. 159–179.

Audebert, A. and Fofana, M. (2009) 'Rice yield gap due to iron toxicity in West Africa', *Journal of Agronomy and Crop Science*, vol. 195, no. 1, pp. 66–76.

Balasubramanian, V., Sié, M., Hijmans, R. J. and Otsuka, K. (2007) 'Increasing rice production in sub-Saharan Africa: challenges and opportunities', *Advances in Agronomy*, vol. 94, no. 1, pp. 55–133.

Becker, M. and Asch, F. (2005) 'Iron toxicity in rice-conditions and management concepts', *Journal of Plant Nutrition and Soil Science*, vol. 168, no. 4, pp. 558–573.

Becker, M. and Johnson, D. E. (2001) 'Improved water control and crop management effects on lowland rice productivity in West Africa', *Nutrient Cycling in Agroecosystems*, vol. 59, no. 2, pp. 119–127.

Biasutti, M., Held, I. M., Sobel, A. H. and Giannini, A. (2008) 'SST forcings and Sahel rainfall variability in simulations of the twentieth and twenty-first centuries', *Journal of Climate*, vol. 21, no. 14, pp. 3471–3486.

Brautigam, D. (1992) 'Land rights and agricultural development in West Africa – a case study of 2 Chinese projects', *Journal of Developing Areas*, vol. 27, no. 1, pp. 21–32.

Brockington, D. (2007) 'Forests, community conservation, and local government performance: the village forest reserves of Tanzania', *Society & Natural Resources*, vol. 20, no. 9, pp. 835–848.

Buri, M. M., Masunaga, T. and Wakatsuki, T. (2000) 'Sulfur and zinc levels as limiting factors to rice production in West Africa lowlands', *Geoderma*, vol. 94, no. 1, pp. 23–42.

Chabi, A., Oloukoi, J., Mama, V. J. and Kiepe, P. (2010) 'Inventory by remote sensing of inland valleys agro-ecosystems in central Benin', *Cahiers Agricultures*, vol. 19, no. 6, pp. 446–453.

Christensen, J. H., Hewitson, B., Busuioc, A., Chen, A., Gao, X., Held, I., Jones, R., Kolli, R. K., Kwon, W. T., Laprise, R., Magaña Rueda, V., Mearns, L., Menéndez, C. G., Räisänen, J., Rinke, A., Sarr, A. and Whetton, P. (2007) 'Regional climate projections', in S. Solomon, D. Qin, M. Manning, Z. Chen, M. Marquis, K. B. Averyt, M. Tignor and H. L. Miller (eds) *Climate Change 2007: The Physical Science Basis. Contribution of Working Group I to the Fourth Assessment Report of the Intergovernmental Panel on Climate Change*, Cambridge University Press, Cambridge and New York, pp. 847–940.

Cook, K. H. and Vizy, E. K. (2006) 'Coupled model simulations of the West African monsoon system: twentieth- and twenty-first-century simulations', *Journal of Climate*, vol. 19, no. 15, pp. 3681–3703.

Dalton, T. J. and Guei, R. G. (2003) 'Productivity gains from rice genetic enhancements in West Africa: countries and ecologies', *World Development*, vol. 31, no. 2, pp. 359–374.

Defoer, T., Wopereis, M. C. S., Idinoba, P. A., Kadisha, T. K. L., Diack, S. and Gaye, M. (2004) *Curriculum for participatory learning and action research (PLAR) for integrated rice management (IRM) in inland valleys of sub-Saharan Africa: facilitators' manual*, WARDA, Cotonou, Benin/IFDC, Muscle Shoals.

Del Amo-Rodriguez, S., Vergara-Tenorio, M. D., Ramos-Prado, J. M. and Porter-Bolland, L. (2010) 'Community landscape planning for rural areas: a model for biocultural resource management', *Society & Natural Resources*, vol. 23, no. 5, pp. 436–450.

Dixon, A. B. and Wood, A. P. (2003) 'Wetland cultivation and hydrological management in eastern Africa: matching community and hydrological needs through sustainable wetland use', *Natural Resources Forum*, vol. 27, no. 2, pp. 117–129.

Djagba, J. F., Rodenburg, J., Zwart, S. J., Houndagba, C. J. and Kiepe, P. (2012) 'Failure and success factors of irrigation system developments in West Africa: a case study from the Ouémé valley in Benin', *Irrigation and Drainage* (under review).

Dries, I. (1991) 'Development of wetlands in Sierra-Leone: farmers rationality opposed to government policy', *Landscape and Urban Planning*, vol. 20, iss 1–3, pp. 223–229.

Dugan, P. J. (1990) *Wetland conservation: a review of current issues and action*, IUCN, Gland.

Erwin, K. (2009) 'Wetlands and global climate change: the role of wetland restoration in a changing world', *Wetlands Ecology and Management*, vol. 17, no. 1, pp. 71–84.

Fagbami, A., Ajayi, S. O. and Ali, E. M. (1985) 'Nutrient distribution in the basement-complex soils of the tropical, dry rainforest of south-western Nigeria 1. Macronutrients – Calcium, Magnesium, and Potassium', *Soil Science*, vol. 139, no. 5, pp. 431–436.

FAO (2010) 'FAOSTAT', http://faostat.fao.org (accessed 7 October 2010).

Fu, R. H. Y., Abe, S. S., Wakatsuki, T. and Maruyama, M. (2010) 'Traditional farmer-managed irrigation system in central Nigeria', *Japan Agricultural Research Quarterly*, vol. 44, no. 1, pp. 53–60.

Giannini, A., Biasutti, M., Held, I. M. and Sobel, A. H. (2008) 'A global perspective on African climate', *Climatic Change*, vol. 90, no. 4, pp. 359–383.

Gumma, M., Thenkabail, P. S., Fujii, H. and Namara, R. (2009) 'Spatial models for selecting the most suitable areas of rice cultivation in the Inland Valley Wetlands of Ghana using remote sensing and geographic information systems', *Journal of Applied Remote Sensing*, vol. 3, no. 1.

Hoerling, M., Hurrell, J., Eischeid, J. and Phillips, A. (2006) 'Detection and attribution of twentieth-century northern and southern African rainfall change', *Journal of Climate*, vol. 19, no. 16, pp. 3989–4008.

Issaka, R. N., Ishida, F., Kubota, D. and Wakatsuki, T. (1997) 'Geographical distribution of selected soil fertility parameters of inland valleys in West Africa', *Geoderma*, vol. 75, nos 1–2, pp. 99–116.

Issaka, R. N., Masunaga, T., Kosaki, T. and Wakatsuki, T. (1996) 'Soils of inland valleys of West Africa general fertility parameters', *Soil Science and Plant Nutrition*, vol. 42, pp. 71–80.

Johnson, D. E. (1997) *Weeds of rice in West Africa*, WARDA, Bouaké.

Kent, R. J. and Johnson, D. E. (2001) 'Influence of flood depth and duration on growth of lowland rice weeds, Côte d'Ivoire', *Crop Protection*, vol. 20, no. 8, pp. 691–694.

Kent, R. J., Johnson, D. E. and Becker, M. (2001) 'The influences of cropping system on weed communities of rice in Côte d'Ivoire, West Africa', *Agriculture, Ecosystems and Environment*, vol. 87, no. 3, pp. 299–307.

Kiepe, P. (2006) *Characterization of three key environments for integrated irrigation-aquaculture and their local names, Integrated Irrigation and Aquaculture in West Africa, Concepts, Practices and Potential*, FAO, Rome, pp. 41–45.

Kristensen, M. K. and Lykke, A. M. (2003) 'Informant-based valuation of use and conservation preferences of savanna trees in Burkina Faso', *Economic Botany*, vol. 57, no. 2, pp. 203–217.

Kyuma, K. (1985) *Fundamental characteristics of wetland soils: characterization, classification and utilization*, International Rice Research Institute (IRRI), Los Banos, Philippines, pp. 193–205.

Leach, M. (1991) 'Engendered environments: understanding natural resource management in the West African forest zone', *IDS Bulletin*, no. 22, pp. 17–24.

Legoupil, J. C., Lidon, B., Blanchet, F. and Jamin, J. Y. (2001) *Mise en valeur et aménagement des bas-fonds d'Afrique de l'Ouest, proposition d'un outil d'aide à l'aménagement: le diagnostic rapide de pré-aménagement (DIARPA)*, CIRAD & IVC/CBF, Montpellier/Bouaké.

Lidon, B., Legoupil, J. C., Blanchet, F., Simpara, M. and Sanogo, I. (1998) 'Le diagnostic rapide de pré-aménagement (DIARPA)', *Agriculture et Développement*, vol. 20, pp. 61–80.

Maconachie, R. (2008) 'New agricultural frontiers in post-conflict Sierra Leone? Exploring institutional challenges for wetland management in the Eastern Province', *Journal of Modern African Studies*, vol. 46, iss. 2, pp. 235–266.

Madge, C. (1995) 'Ethnography and agroforestry research: a case study from the Gambia', *Agroforestry Systems*, vol. 32, no. 2, pp. 127–146.

McCartney, M. P. and Houghton-Carr, H. A. (2009) 'Working wetland potential: an index to guide the sustainable development of African wetlands', *Natural Resources Forum*, vol. 33, no. 2, pp. 99–110.

Narteh, L. T. and Sahrawat, K. L. (1999) 'Influence of flooding on electrochemical and chemical properties of West African soils', *Geoderma*, vol. 87, iss 3–4, pp. 179–207.

Ogban, P. I. and Babalola, O. (2003) 'Soil characteristics and constraints to crop production in inland valley bottoms in south-western Nigeria', *Agricultural Water Management*, vol. 61, no. 1, pp. 13–28.

Oladele, O. I., Bam, R. K., Buri, M. M. and Wakatsuki, T. (2010) 'Missing prerequisites for Green Revolution in Africa: lessons and challenges of Sawah rice eco-technology development and dissemination in Nigeria and Ghana', *Journal of Food Agriculture & Environment*, vol. 8, no. 2, pp. 1014–1018.

Oosterbaan, R. J., Gunneweg, H. A. and Huizing, A. (1987) 'Water control for rice cultivation in small valleys of West Africa', *Annual Report ILRI*, Wageningen, pp. 30–49.

Ouédraogo, O., Neumann, U., Raynal Roques, A., Sallé, G., Tuquet, C. and Dembélé, B. (1999) 'New insights concerning the ecology and the biology of *Rhamphicarpa fistulosa* (Scrophulariaceae)', *Weed Research*, vol. 39, no. 2, pp. 159–169.

Perfecto, I. and Vandermeer, J. (2008) 'Biodiversity conservation in tropical agroecosystems: a new conservation paradigm', *Annals of the New York Academy of Sciences*, vol. 1134, no. 1, pp. 173–200.

Raj, P. A. (1995) 'Multicriteria methods in river basin planning: a case study', *Water Science and Technology*, vol. 31, no. 8, pp. 261–272.

Raunet, M. (1985) 'Bas-fonds et riziculture en Afrique: approche structurale comparative', *Agronomie Tropicale*, vol. 40, no. 3, pp. 181–201.

Rodenburg, J. and Demont, M. (2009) 'Potential of herbicide resistant rice technologies for sub-Saharan Africa', *AgBioForum*, vol. 12, nos 3–4, pp. 313–325.

Rodenburg, J. and Johnson, D. E. (2009) 'Weed management in rice-based cropping systems in Africa', *Advances in Agronomy*, vol. 103, pp. 149–218.

Rodenburg, J., Riches, C. R. and Kayeke, J. M. (2010) 'Addressing current and future problems of parasitic weeds in rice', *Crop Protection*, vol. 29, no. 3, pp. 210–221.

Rodenburg, J., Both, J., Heitkönig, I. M. A., Koppen, C. S. A., Sinsin, B., Van Mele, P. and Kiepe, P. (2012) 'Land-use and biodiversity in unprotected landscapes: the case of non-cultivated plant use and management by rural communities in Benin and Togo', *Society & Natural Resources*, vol. 25, iss. 12, pp. 1221–1240.

Rodenburg, J., Saito, K., Kakai, R. G., Toure, A., Mariko, M. and Kiepe, P. (2009) 'Weed competitiveness of the lowland rice varieties of NERICA in the southern Guinea Savanna', *Field Crops Research*, vol. 114, no. 3, pp. 411–418.

Rodenburg, J., Zossou-Kouderin, N., Gbèhounou, G., Ahanchede, A., Touré, A., Kyalo, G. and Kiepe, P. (2011) '*Rhamphicarpa fistulosa*, a parasitic weed threatening rain-fed lowland rice production in sub-Saharan Africa: a case study from Benin', *Crop Protection*, vol. 30, no. 10, pp. 1306–1314.

Saito, K., Azoma, K. and Sokei, Y. (2010) 'Genotypic adaptation of rice to lowland hydrology in West Africa', *Field Crops Research*, vol. 119, no. 2, pp. 290–298.

Sakurai, T. (2006) 'Intensification of rainfed lowland rice production in West Africa: present status and potential Green Revolution', *Development Economics*, vol. 44, no. 2, pp. 232–251.

Shepherd, G. (1992) *Managing Africa's tropical dry forests: a review of indigenous methods*, Overseas Development Institute, London.

Sié, M., Séré, Y., Sanyang, S., Narteh, L. T., Dogbe, S., Coulibaly, M. M., Sido, A., Cissé, F., Drammeh, E., Ogunbayo, S. A., Zadji, L., Ndri, B. and Toulou, B. (2008) 'Regional yield evaluation of the interspecific hybrids (*O. glaberrima* × *O. sativa*) and intraspecific (*O. sativa* × *O. sativa*) lowland rice', *Asian Journal of Plant Sciences*, vol. 7, no. 2, pp. 130–139.

Silvius, M. J., Oneka, M. and Verhagen, A. (2000) 'Wetlands: lifeline for people at the edge', *Physics and Chemistry of the Earth Part B-Hydrology Oceans and Atmosphere*, vol. 25, no. 7, pp. 645–652.

Thenkabail, P. S. and Nolte, C. (1996) 'Capabilities of Landsat-5 Thematic Mapper (TM) data in regional mapping and characterization of inland valley agroecosystems in West Africa', *International Journal of Remote Sensing*, vol. 17, no. 8, pp. 1505–1538.

Thiombiano, L., Jamin, J. Y. and Windmeijer, P. N. (1996) *Inland valley development in West Africa: regional state of the art*, Inland Valley Consortium, WARDA, Bouaké.

Toure, A., Becker, M., Johnson, D. E., Kone, B., Kossou, D. K. and Kiepe, P. (2009) 'Response of lowland rice to agronomic management under different hydrological regimes in an inland valley of Ivory Coast', *Field Crops Research*, vol. 114, no. 2, pp. 304–310.

van Duivenbooden, N., Windmeijer, P. N., Andriesse, W. and Fresco, L. O. (1996) 'The integrated transect method as a tool for land use characterisation, with special reference to inland valley agroecosystems in West Africa', *Landscape and Urban Planning*, vol. 34, iss. 2, pp. 143–160.

Wakatsuki, T. and Masunaga, T. (2005) 'Ecological engineering for sustainable food production and the restoration of degraded watersheds in tropics of low pH soils: focus on West Africa', *Soil Science and Plant Nutrition*, vol. 51, no. 5, pp. 629–636.

Warner, J. F. (2006) 'More sustainable participation? Multi-stakeholder platforms for integrated catchment management', *International Journal of Water Resources Development*, vol. 22, no. 1, pp. 15–35.

Windmeijer, P. N. and Andriesse, W. (1993) *Inland valleys in West Africa: an agro-ecological characterization of rice growing environments*, ILRI, Wageningen.

Wopereis, M. C. S. and Defoer, T. (2007) 'Moving methodologies to enhance agricultural productivity of rice-based lowland systems in sub-Saharan Africa', in A. Bationo, B. Waswa, J. Kihara and J. Kimetu (eds) *Advances in Integrated Soil Fertility Management in sub-Saharan Africa: Challenges and Opportunities*, Springer, the Netherlands, pp. 1077–1091.

Wopereis, M. C. S., Defoer, T., Idinoba, M. E., Diack, S. and Dugué, M. J. (2007) *Participatory learning and action research (PLAR) for integrated rice management (IRM) in inland valleys of sub-Saharan Africa: technical manual*, WARDA, Cotonou, Benin/IFDC, Muscle Shoals.

Wyman, M. and Stein, T. (2010) 'Examining the linkages between community benefits, place-based meanings, and Conservation Program involvement: a study within the Community Baboon Sanctuary, Belize', *Society & Natural Resources*, vol. 23, no. 6, pp. 542–556.

11 Conclusions

Transforming wetland livelihoods

Adrian Wood, Alan Dixon and Matthew McCartney

Introduction

This final chapter outlines the various themes and issues that have emerged from the preceding chapters, identifying common experience and lessons that can be drawn. This analysis is made primarily with reference to our ideas regarding the evolving discourse of wetland management and the way ahead as presented in Chapter 1, which fundamentally places people at the centre of thinking about wetlands in Africa. This is the direction in which we believe understanding of wetlands must progress if the multiple eco-system services (ESS) and benefits from wetlands are to contribute to, and sustain, both development and environmental goals.

Putting people at the centre of wetland management thinking reminds us that it is human well-being, facilitated through sustainable livelihoods, that should be the central goal of wetland policy-making and practice in Africa. Indeed, the case studies presented throughout this book have high-lighted the considerable range of wetland-based livelihoods that exist, and that make a contribution to the development needs of many of the poorest people across the continent. Of these, the most significant is agriculture. Wetlands remain the new agriculture frontier in Africa (Dixon and Wood, 2003) and hence we argue the need for a much greater focus on wetland agriculture among the whole wetland policy-making community. The case studies in this volume complement the growing body of literature and field evidence (see for example Kangalawe and Liwenga, 2005; Schuyt, 2005; Kipkemboi *et al.*, 2007; Maconachie, 2008; Rebelo *et al.*, 2010; Nabahungu and Visser, 2011; Leauthaud *et al.*, 2012) that serve to justify and emphasize the importance of placing wetland agriculture central in Africa's evolving wetlands discourse. Moreover, with African development policy (and inter-national aid) continuing to be dominated by the Millennium Development Goals (MDGs) and by the growing concerns about increasing food pro-duction, there is little doubt that wetlands – through their various ESS, especially agriculture – have the potential to play a key role in eradicating extreme poverty and hunger (MDG1), ensuring environmental sustain-ability (MDG7) and improving food security. Recognizing more explicitly the people–wetland linkages that exist and facilitating the development of sustainable wetland livelihoods provides an important contribution not only towards achieving the MDGs, but also towards providing a buffer and adaptive strategy for the challenges of population growth, increasing food production and climate change (see Chapters 2, 3 and 8).

To address these challenges, wetland management strategies and policies must be holistic and innovative. Critically they must recognize the importance of the full range of ESS provided by wetlands and also acknowledge and accept that wetland transformation is an inevitable consequence of livelihood development. As we have seen throughout this book (Chapters 3, 8 and 10) it is possible to make trade-offs between different wetland uses (and hence ESS) which can result in economically, environmentally and socially sustainable outcomes, e.g. wetland agriculture developed to some degree alongside the maintenance of other ESS. The ecological character of wetlands will change, but the value of wetlands to society will increase provided transformation does not undermine other ESS (see trade-offs discussion below). Indeed, under the present circumstances, maintaining Africa's wetlands in a pristine condition – itself a nebulous term – and without any ecological change, has become an increasingly less viable option and is no longer morally justifiable.

Key themes emerging

The diverse case studies have drawn on experience from many parts of the continent and from work at different levels, although predominantly from field experience with rural communities. Despite this diversity a number of common themes can be identified.

Contribution of wetland agriculture to livelihoods

The experiences reported in this book make it clear that increasingly wetlands must be seen in a new light, as important assets within peoples' wider livelihood portfolio, and that these assets are capable of providing benefits beyond the ESS they provide in their natural state. To many people, wetlands now constitute a critical agricultural resource that 'adds value' to their livelihood system and can stimulate economic development. Hence, if wetlands are to be utilized sustainably this is something that cannot be ignored; wetland agriculture must become central to policy-making and management.

The current contribution of wetlands to peoples' livelihoods is confirmed in all the chapters presented in this book. However, while a range of different livelihood activities are identified, wetland cultivation arguably emerges as the most important in terms of its contribution to economic and social well-being. The case studies also reveal that wetland livelihood activities are both spatially and temporally variable, and to a large extent are determined by the wider livelihood assets available to individuals, i.e. access to natural, human, financial, social and physical capital (see Figure 1.4). Where, for example, people lack access to upland agricultural land, either permanently or periodically, wetlands may offer a means of producing subsistence crops (Chapters 3 and 5), and hence become part of a survival or coping strategy for many people, especially the poor. Similarly, seasonal climatic conditions may result in an increased reliance on wetland resources during dry periods (Chapters 3 and 4). In some cases, it is the relatively better-off 'asset rich' who are recognizing that wetlands offer an opportunity to supplement their incomes through crop production, and sometime livestock, to accumulate wealth (Chapter 2). Clearly such a diversity of dynamic wetland livelihood situations renders policy-making and management guidance particularly challenging, but as we have seen in several cases (Chapters 5 and 7), the adoption of an adaptive socio-ecological systems (SES) approach (Berkes *et al.*, 2003; Ostrom, 2007) to the management of wetland agriculture offers one potential way forward for ensuring local sensitivity and sustainability. We also note here the important role that local institutional arrangements need to play in developing the rules of agricultural engagement with wetlands, and promoting effective management, sustainable use and equitable access (Chapters 3, 4 and 5).

Pressures on wetlands

It is clear from the case studies that the livelihood demands on wetlands will grow in the foreseeable future as populations increase (Chapter 8), urbanization creates new market opportunities (Chapter 10) and upland degradation most likely continues (Chapter 3). Moreover, as predicted by Boko *et al.* (2007), climate change is likely to exacerbate food shortages and land degradation, and result in more permanent and seasonal migration in search of livelihood opportunities and often agricultural resources. At the national level, government policies that aim to increase domestic food security and reduce reliance on food imports are likely to drive the search for more agricultural land, and the intensified use of all land. Hence, the wetland agricultural frontier is likely to continue to expand for some decades, while the pressure on existing rain-fed/upland farmland will increase. In addition, irrigation development will expand, often in wetlands (Chapters 8 and 9). The socio-economic implications of these developments at the local level are likely to be increasing competition (and potentially conflict) over land and water in wetlands (Chapters 8 and 9), many of which exist as common property resources (CPRs). As a result, it is clear that there is a need for wetland-using communities to adapt their wetland management strategies to this range of pressures, to ensure sustainable and equitable use.

In the absence of the development of adaptive wetland management strategies, however, the environmental implications of increasing pressure on wetlands are clear – growing demands on wetlands for food, water and other provisioning services can lead to wetland degradation in the form of gulley erosion, over drainage, desiccation and excessive flooding (Chapters 3 and 4). While some technical solutions are possible and some have been discussed in the various chapters (notably Chapters 3, 6 and 9), clearly building up a comprehensive understanding of how specific wetlands function (that draws upon both 'local' and 'scientific' knowledge) should be a fundamental prerequisite to any development of provisioning services so as to better manage pressures on wetlands and prevent degradation.

Resilience and trade-offs

Although recognized as a widespread phenomenon throughout the litera-ture, wetland degradation does not emerge as a dominant theme from this book. Rather we have sought to draw attention to the many cases of long-term and sustainable use of wetlands for agriculture that have been, or can be, achieved. In many cases a trade-off in terms of altering the balance of ESS within wetlands has been achieved, which allows the use of wetlands for provisioning services, with sufficient other ESS maintained, albeit in a reduced manner, to ensure this new wetland use and balance of ESS is sustainable. This experience suggests that some wetlands are more resilient than others and that provided the extent and degree of transformation are kept within limits, and specific damaging practices are avoided, some transformation of wetlands for agriculture without degradation is possible in community-managed wetlands. In this book several studies show that sus-tainable wetland agriculture is possible (Chapters 2, 3, 4, 6 and 10), while in Chapter 6, which assesses the sustainability of wetland cultivation, it is shown that there are only 'medium' levels of environmental risk in the specific cases studied, and that is before any adjustment of management practices to improve sustainability. Further, specific measures can be identified that reduce the risks of degradation, especially through land management prac-tices in the wetlands and catchments, the choice of crops and patterns of land use, river regulation and flow re-optimization of the discharge from dams, and catchment-wide coordination of management (Chapters 3, 5, 9 and 10). While these measures can help to ensure the timely and neces-sary availability of water and its retention for wetland agriculture, questions remain on the extent and degree of transformation which is acceptable in wetlands whilst sustaining cultivation and how such guidance must be adjusted to different situations.

Local knowledge, local institutions

A recurrent theme through the case studies is the level of local knowledge that wetland-using communities have developed and the potential of this for the further development of sustainable-use regimes. Local knowledge is based on regular observations and experiences, and, in this sense, it can be seen as local science that can complement 'external', 'scientific' knowledge and enhance the efficacy of management interventions. Although critics of local knowledge have pointed to its unreliability and subjectivity, proponents suggest that this somewhat misses the point of applying an interpretative approach to SES. Identifying local knowledge is not so much a quest for 'fact' but rather an attempt to understand the meaning and significance that local people place on their environment, which subsequently influences their interaction with resources. One key conclusion that can be drawn from the studies presented in this book, therefore, is that whilst wetlands have a different meaning for different people (local and external) in both time and space, each has a part to play in building up a composite, interdisciplinary understanding of reality that can contribute to the development of sustainable wetland management to ensure livelihood benefits in the long term.

While the importance of local knowledge and local institutions for wetland management are addressed explicitly in Chapter 4, several chapters raise this as an issue needing more attention. A recurring theme is the need for the coordinated management of water resources at both catchment and wetland levels (Chapters 3, 5, 8 and 10), and the coordination of land use within wetlands themselves (Chapters 3, 5 and 10). Here, there is evidence that local, community-based institutions can play an important role in coordinating land and water resource management through collective action, which can support sustainable wetland use. Providing they fulfil certain criteria (see Ostrom, 1990) such CPR institutions can reduce the risk of a 'tragedy of the commons' type situation (see Hardin, 1968), and play a role in balancing ESS to sustain livelihood benefits from wetlands. These CPR institutions can also be a way in which multiple stakeholders with different socio-economic interests in wetlands can be engaged and conflicts resolved. However, it is difficult to predict at what scale they operate most effectively and the issue of CPR institutions for wetlands is certainly an area where the need for further study is identified.

Government policies

The interaction between wetland-using communities and government is clearly an important issue affecting wetland management as several studies have shown. The areas of government interaction are diverse, ranging from food security policies (Chapter 4), to tenure and security of access to land (Chapter 5), as well as national or international wetland policies (Chapter 7). In addition, there is the issue of the nature of the relationship between government and people including whether there is communication between the parties or a top-down planning approach, and also the role of traditional authorities in the relations between communities and central government. Government action can also support or undermine local institutions for wetland management and the effectiveness of their work. In particular, government policy can constitute a major direct or indirect driver of pressure on wetlands, and hence can fundamentally shape local peoples' relation with wetlands. For example, land tenure and the nature of agrarian relations, which result from government policies, can play a fundamental role in determining access to resources, whilst national wetland policy itself (where it exists) and the nature of its implementation can be critical in achieving sustainable wetland use, or not.

In many countries, national wetland policies continue to be dominated by a conservation agenda that focuses on larger wetlands of international importance, at the expense of smaller wetlands that contribute to peoples livelihoods. In the absence of specific policies that recognize the interrelated social, economic and environmental significance of these smaller wetlands, national food security or agricultural policies continue to take a very uniform or generalizing view that often threatens the sustainability of wetlands and wetland-based livelihoods. As a result, community sensitive development of government policy is needed that recognizes the multiple roles of wetlands and their dynamics.

Policies and communities

The over-arching issue here is one of the policy-making process itself, which on the evidence presented in the case studies, remains predominantly 'top-down' in nature and insensitive to local conditions and local institutional arrangements. Little space is provided for wetland communities to engage in these policy debates, and for the outputs to be developed by them, reflect their views and be owned by them. Despite almost 30 years of the 'farmer first' paradigm (Chambers, 1983; Chambers *et al.*, 1989), rarely are local wetland-users consulted or included in policy-making, hence top-down policies override local knowledge, management systems and site-specific understanding of wetlands. However, many of the chapters here provide examples of how this situation is gradually being acknowledged and addressed, either through governments seeking to facilitate more 'grassroots' decision making (Chapter 7) or acknowledging the mistakes of the past (Chapter 9), or via non-governmental organizations (NGOs) becoming active in facilitating community-based adaptive management approaches (Chapter 5) and empowering local institutional arrangements (Chapters 3 and 4). Nonetheless, there is much to do to ensure that the interests of communities are fully considered in policy development, and that communities are engaged so that they understand the wider basin-level issues that government policies have to address.

From lessons to action

In view of the lessons that have emerged from the case studies, we suggest a number of specific actions that can potentially inform and improve wetland management in Africa so that it can support the development of sustainable livelihoods and contribute towards the future challenges faced in African development. However, care must be taken in making such recommendations since, as discussed below, wetlands and their circumstances are spatially and temporally diverse, and require adaptive approaches. The following action points are developed from the above discussion of lessons, but they do not provide a one-to-one response. Rather, the focus is on identifying specific measures or approaches that can together contribute to sustainable wetland management, and these are often relevant to more than one of the lessons above.

Adopting a socio-ecological systemic approach

A systemic approach, as embodied in SES thinking, must be the starting point for maintaining wetland functioning. This recognizes the essential interrelationships between communities and their environments, and makes it clear that people must be the starting point for any wetland management intervention. This linkage has been recognized in all the chapters in this book where there has been a focus on the dynamic relationship between wetland-users, their immediate wetland and the wider catchment (see for example, the Functional Landscape Approach presented in Chapter 3). Furthermore, the systemic linkages extend beyond the catchment to government policies, food prices and urban demands that are some of the influences on wetland use and management (Chapter 5). This raises the fundamental issue of scale and the need for wetland management to consider and incorporate a sensitivity to the local–national, catchment–wetland, individual–community linkages, processes and flows (Figure 11.1). As a result, it is clear that the SES approach requires a rethinking of the functional relationship between wetlands, people and livelihoods. It suggests that we must look beyond the environmental discourse that has dominated past discussions, and focus much more on the way communities adapt and manage their resources and assets, and especially how they manipulate their environmental resources to achieve sustainable and resilient livelihoods.

Reflexive and learning approach to adaptive wetland management

Wetland management must involve a continual learning process (see Chapter 5). As has been shown in most cases highlighted in this book, there is a need for increased understanding of the way the environmental, economic, social and political processes interact with wetlands and their communities and affect sustainable management. Communities who

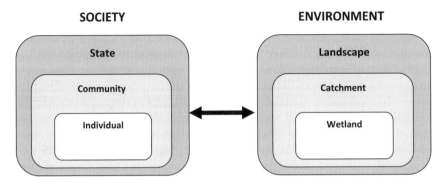

Figure 11.1 Wetland management for sustainable livelihoods means considering the links between society and the environment at different scales

manage wetlands, or projects that are engaged in supporting such management, must also reflect on their own interventions and management practices and be willing to adapt their actions and their recommendations in a continual, iterative learning process. Flexibility is needed in managing these resources, especially given the increasing pressures identified above, while management must be adapted to each site's specific circumstances.

Incentives and understanding peoples' motivations for wetland use

Putting people at the centre of wetland thinking brings to the fore a neglected aspect of wetland management – the decision making and motivations of wetland-users. Demonstrable income or livelihood benefits from wetlands can generate appreciation of the value of these areas and provide an incentive for people to manage them in a sustainable manner in order to maintain these benefits. To be self-sustaining these incentives should originate from market opportunities, which are most likely to be available from produce that can be sold, although some payments for environmental services related to flood amelioration or carbon sequestration may become increasingly significant for some areas in the future. Again, we assert here that past approaches prohibiting the use of wetlands as the way to achieve their sustainable management have largely failed because they have marginalized local people rather than giving them an incentive to become actively involved in wetland management and generate benefits from these areas. Adding value to resources is arguably a much more effective way of motivating people to engage in resource management practices that encourage sustainable use, providing that the rules of engagement and access are developed and agreed at the community level to prevent overexploitation and degradation.

Trade-offs

As with any human endeavour, there are trade-offs associated with wetland management especially when it involves the development of agriculture. Many ESS are interdependent and so altering the system to enhance one (e.g. food production) will have a knock-on effect (positive or negative) on others. Inappropriate wetland agriculture can lead to degradation and the complete loss of some ESS, ultimately undermining those services that support the agriculture itself (Wood and van Halsema, 2008). Consequently, sustainability requires that wetlands are managed for multiple ESS and livelihood strategies. Although in some instances this may mean greater emphasis is placed on conservation to protect key functions, managing wetlands for livelihoods through agricultural development is not congruent with managing them solely to 'protect' wetland ecosystems. In reality there will often be trade-offs between livelihood and purely conservation requirements that require careful management. The objective of addressing these trade-offs should not be to maximize values for conservation and livelihood development simultaneously but rather to produce maximum net benefits for people whilst at the same time avoiding fundamental ecological threats (Sellamuttu *et al.*, 2008). Hence, an approach is required that increases the overall productivity of a wetland whilst ensuring livelihoods and food security are enhanced rather than harmed by agricultural development.

The challenge of managing wetlands for people and ESS is that the biophysical processes that underpin the services are often not well understood, making it difficult to predict what the exact implications of anthropogenic change to a wetland in a specific situation will be. Furthermore, it is often the cumulative effect of managing several wetlands within a landscape that has to be considered when seeking to maximize net benefits. It is likely that future research will shed light on the different services provided by different wetlands and how these are influenced by the context of the landscape in which they are situated. However, it is not possible to wait for this research to be completed. The reality is that wetland agriculture is increasingly prevalent and, as discussed above, a variety of drivers mean that it is likely to increase rapidly in the near future. Consequently, management decisions will have to be made in a context of high uncertainty and with only limited understanding. Recognizing this, the need for adaptive management approaches is clearly confirmed again.

Empower local institutional arrangements for wetland management

The management of wetlands should be undertaken by those whose livelihoods depend upon them and who have first-hand experience of the socio-ecological dynamics of the system, rather than by state institutions or high-level agencies that are unable to adapt policies to specific conditions.

Local community-based institutions have been shown to be effective in managing natural resources sustainably, since they draw upon local site-specific knowledge, and in many cases these institutions are developed from a shared concern (social capital) for resource and livelihood sustainability. A key challenge, however, lies in facilitating the creation of these institutions which, in order to function effectively and be sustainable in the long term, should have representation from all wetland stakeholders to enable the development of mutually agreed rules and regulations for wetland access and management. Various factors will affect the development of effective local institutions, including the size and diversity of the community and interest groups represented, as well as the local government situation. Critically, the external policy environment should facilitate, rather than undermine, the functioning of these institutions at the local level. In addition, as discussed in Chapter 4 in particular, NGOs can potentially play a key role here in supporting the grassroots development of these institutions. Local institutions must also link to wider basin-level management institutions, and this interaction will influence effective management at specific sites, as well as the basin as a whole.

Policies

Many areas of government policy can affect wetlands, and not just wetland policies. As discussed above, policy development is often insensitive to the diverse livelihoods that communities obtain from wetlands, with either overly conservationist policies or demands to intensively develop wetlands that will undermine their sustainable use. In order to achieve a policy environment that will support sustainable wetland management there is a need for more engagement with wetland stakeholders, especially wetland-using communities, when wetland policies are developed. This could involve the use of wetland management groups or institutions who can represent the users of wetlands. Clearly there are many levels at which trade-offs have to be made, between different stakeholder/interest groups, between local or site-specific interests and basin-level concerns, and between basin-level interests and national ones. The only way to address these different interests is by greater involvement of those who use them, the stakeholders.

Demonstration and evolving best practices

Lessons must be learned from cases where wetlands have been managed sustainably. Not only do these need to be documented, but there is a need to use these as demonstration sites that can help spread good practice. In particular, techniques of land and water management that have proved successful need to be disseminated, recognizing that they will have probably benefitted from specific socio-economic and biophysical situations. A list of good practices that does not refer to the specific circumstances where success has been achieved will be more dangerous than helpful. Techniques and practices are only part of what is needed to achieve sustainable wetland management as organizational, economic and political factors are all equally important (see Chapter 1). Such an understanding must not just be at the field level, but also inform the policies developed to support wetland management.

Further learning

In our exploration of the management of wetlands for sustainable livelihoods in Africa, several knowledge gaps have been identified that must be addressed if appropriate management regimes and approaches are to be developed.

Detailed interdisciplinary research

While the chapters in this book have tried to bring together material from a range of disciplines, it is clear that there remain major divisions in approaches and analysis amongst people working on wetlands which has resulted in assumptions and even guesswork in attempts to bridge existing gaps in knowledge. What is needed to take forward our understanding of how to achieve sustainable wetland management is more interdisciplinary approaches to wetlands. By this we mean 'joined-up' thinking that goes beyond a research programme where one person researches soils, another interviews farmers and where neither talks to each other. Rather we need to have an integration of research around a common understanding in which the different disciplinary contributions are incorporated. This is the way to avoid assumptions about cause and effect. Detailed investigations that involve the local communities as the focus are the best way to take this forward, rather than generalized studies, and so we would stress the need for a concurrent methodological paradigm shift towards intensive and qualitative approaches.

Transformation, trade-offs and limits of change

The extent to which wetland transformation can be pursued without the functioning of these areas being undermined so that their value to society is reduced rather than enhanced, is still poorly understood. The Millennium Ecosystem Assessment (MA) and the Guidelines on Agriculture and Wetlands Interaction (GAWI) studies discuss the balance of ESS that needs to be maintained but they fail to define what this means in practice (Wood and van Halsema, 2008). At the same time, there is, in this book and elsewhere, growing evidence that wetlands can be transformed for agriculture and have their ecological character altered, while the overall contributions of their ESS for society can remain of value. In fact, many regulating services will continue even if a wetland is moderately modified. Less clear, however, are the thresholds of change before the ESS are lost. Indentifying this dynamic process of change and the point at which transformation trade-off becomes critical and leads to the loss of ESS is a major challenge. This seriously complicates achieving management that will ensure sustainable use. We suggest that research is needed to better understand this dynamic and that this will require community-based but externally supported research.

Rethinking wetlands for sustainable livelihoods in Africa

In this book we have sought to reconceptualize wetlands as natural resources within the discourse of sustainable development. Wetlands are vital assets for many households and communities, and understanding that viewpoint is the critical change in wetland thinking we see as necessary for ensuring wetlands make a full contribution to the future challenges faced from population growth, climate change and poverty reduction. We believe it is essential to reposition the wider wetland management debate away from an environmental management and conservation focus with development outcomes added on, to one of livelihood development that is based on the sustainable use of these resources with positive environmental outcomes, in terms of maintaining sufficient ESS to ensure the sustainable provisioning services.

We also present a different view of wetlands recognizing that for most wetlands some change in their ecological character is inevitable and can be acceptable in terms of maintaining the resource base and ESS, provided the transformation and trade-off processes are managed carefully. Our view of livelihood development and resilience in Africa is one with small inland wetlands, mostly seasonal wetlands, transformed to varying degrees and managed by active, empowered communities with local institutional arrangements supported by a national policy environment. In this way the managed transformation of wetlands will enhance and improve the security of livelihoods for rural communities and hence ensure wetlands contribute to the various MDGs where they have a valuable role to play.

In the coming decades, wetlands in Africa will be increasingly used by people and relied upon for multiple benefits, of an economic, social and environmental nature. It is essential that they are used sustainably so that they are maintained in a state where they can help address the challenges for human development and livelihood security facing Africa throughout the twenty-first century.

References

Berkes, F., Colding, J. and Folke, C. (eds) (2003) *Navigating Social-ecological Systems: Building Resilience for Complexity and Change*, Cambridge University Press, Cambridge.

Boko, M., Niang, I., Nyong, A., Vogel, C., Githeko, A. and Medany, M. (2007) 'Africa', in M. L. Parry, O. F. Canziani, J. P. Palutikof, P. J. Linden and C. E. van der Hanson (eds) *Climate Change 2007: Impacts, Adaptation, Vulnerability: Contribution of Working Group II to the Fourth Assessment Report of the Intergovernmental Panel on Climate Change*, Cambridge University Press, Cambridge.

Chambers, R. (1983) *Rural Development: Putting the Last First*, Longman, London.

Chambers, R., Pacey, A. and Thrupp, L. A. (1989) *Farmer First: Farmer Innovation and Agricultural Research*, Intermediate Technology Publications, London.

Dixon, A. B. and Wood, A. P. (2003) 'Wetland cultivation and hydrological management in eastern Africa: Matching community and hydrological needs through sustainable wetland use', *Natural Resources Forum*, vol. 27, no. 2, pp. 117–129.

Hardin, G. (1968) 'The tragedy of the commons', *Science*, vol. 162, pp. 1243–1248.

Kangalawe, R. Y. M. and Liwenga, E. T. (2005) 'Livelihoods in the wetlands of Kilombero Valley in Tanzania: Opportunities and challenges to integrated water resource management', *Physics and Chemistry of the Earth*, vol. 30, pp. 968–975.

Kipkemboi, J., van Dam, A. A., Mathooko, J. M. and Denny, P. (2007) 'Hydrology and the functioning of seasonal wetland aquaculture–agriculture systems (Fingerponds) at the shores of Lake Victoria, Kenya', *Aquacultural Engineering*, vol. 37, no. 3, pp. 202–214.

Leauthaud, C., Duvail, S., Hamerlynck, O., Paul, J., Cochet, H., Nyunja, J., Albergel, J. and Grunberger, O. (2012) 'Floods and livelihoods: The impact of changing water resources on wetland agro-ecological production systems in the Tana River Delta, Kenya, *Global Environmental Change*, in press.

Maconachie, R. (2008) 'New agricultural frontiers in postconflict Sierra Leone? Exploring institutional challenges for wetland management in the Eastern Province', *The Journal of Modern African Studies*, vol. 46, no. 2, pp. 235–266.

Nabahungu, N. L. and Visser, S. M. (2011) 'Contribution of wetland agriculture to farmers' livelihood in Rwanda', *Ecological Economics*, vol. 71, pp. 4–12.

Ostrom, E. (1990) *Governing the Commons: The Evolution of Institutions for Collective Action*, Cambridge University Press, Cambridge.

Ostrom, E. (2007) Sustainable social-ecological systems: an impossibility?, http://www.indiana.edu/~workshop/publications/materials/conference_papers/W07-2_Ostrom_DLC.pdf (accessed 2nd August 2012).

Releblo, L.-M., McCartney, M. P. and Finlayson, C. M (2010) 'Wetlands of sub-Saharan Africa: Distribution and contribution of agriculture to livelihoods', *Wetlands Ecology and Management*, vol. 18, no. 5, pp. 557–572.

Schuyt, K. (2005) 'Economic consequences of wetland degradation for local populations in Africa', *Ecological Economics*, vol. 53, pp. 177–190.

Sellamuttu, S. S., de Silva, S., Nguyen Khoa, S. and Samarakoon, J. (2008) *Good Practices and Lessons Learned in Integrating Ecosystem Conservation and Poverty Reduction Objectives in Wetlands*, IWMI, Colombo.

Wood, A. P. and van Halsema, G. E. (2008) *Scoping Agriculture-Wetland Interactions: Towards a Sustainable Multiple Response Strategy*, FAO, Rome.

Index